ESSAY

ON

SHEEP;

THEIR VARIETIES—ACCOUNT OF THE MERINOES OF SPAIN, FRANCE, &c.

REFLECTIONS ON THE BEST METHOD OF TREATING THEM, AND RAISING A FLOCK IN THE UNITED STATES;

TOGETHER WITH

MISCELLANEOUS REMARKS

ON

SHEEP AND WOOLLEN MANUFACTURES.

BY ROBERT R. LIVINGSTON, LL. D.

President of the Society for the Promotion of Useful Arts, Member of the American Philosophical Society, President of the American Society of Fine Arts, Corresponding Member of the Agricultural Society of the Seine, Honorary Member of the Agricultural Society of Dutchess County.

Printed by Order of the Legislature of the State of New-York.

NEW-YORK:

PRINTED BY T. AND J. SWORDS,

No. 160 Pearl-Street.

1809.

Originally published 1809
Reprinted 1994 by
United States Historical Research Service
Schenectady, New York
ISBN 0-9633659-1-6

#14117-$24.95

The Remarkable Man Who Brought Merino Sheep to America

Robert R. Livingston

(1746-1813)

The *Essay on Sheep* (1809) is a prized classic both for the historian and for today's practical sheepman. Its author, who financed Robert Fulton's steamboat and became the first president of the New York Agricultural Society, like Washington and many other of America's first governmental leaders, was active in several fields other than politics. But as a jurist and statesman, Chancellor Livingston:

- was a member of the Continental Congress
- helped draft the Declaration of Independence
- was the first Secretary of Foreign Affairs
- administered the oath of office to George Washington
- and negotiated the 1803 purchase of the Louisiana Territory while Jefferson's ambassador to France.

Researched in Europe and then tested on the ancestral 160,000 acre Livingston Manor in Dutchess and Columbia Counties of New York State, Robert Livingston's *An Essay on Sheep* is a priceless treasure of information about daily life in Early America. It is also in a sense a reminder of what constitutes true patriotism.

State of New-York.

In Senate, March 3d, 1809.

WHEREAS the Honourable ROBERT R. LIVINGSTON has laid before the Society for the Promotion of Useful Arts, a manuscript Treatise on Sheep, which the Society judge eminently calculated to diffuse general information as to the best mode of raising and managing the Merino breed of Sheep, from his own experience, and from the best treatises on that subject; and whereas, in the opinion of the said Society, the agricultural interest of this State would be greatly promoted by the printing and circulating the said manuscript—

Therefore Resolved, (if the Honourable the Assembly concur herein) That one thousand copies of the said manuscript be printed in such manner as the said Society shall direct, and that they be distributed in the same manner as the Laws and Journals are by law directed to be distributed, and that each member of the said Society shall also be entitled to one copy; and that the remaining copies shall be at the disposal of the said Society; and that the Legislature will by law make provision for the printing of the said manuscript.

By order of the Senate.

S. VISSCHER, *Clerk.*

In Assembly, March 7th, 1809.

Resolved, That this House do concur with the Honourable the Senate in their preceding resolution.

By order of the Assembly.

D. RODMAN, *Clerk.*

Copy of a Resolution of the Society for the Promotion of Useful Arts.

Resolved, That the Vice-President be requested to transmit to the Honourable ROBERT R. LIVINGSTON, Esq. the concurrent resolutions of the Senate and Assembly for publishing his Treatise on Sheep, and request him to superintend the publication thereof by such Printer as he shall choose; and that he direct five hundred copies, in addition to those ordered by the resolutions, to be published for the Society.

A true copy.

JAMES LOW, *Secretary.*

Albany, *March 8th*, 1809.

PREFACE.

THE hope of acquiring such information in agriculture and the arts as would be useful to my fellow citizens, was not one of my smallest motives for accepting a foreign mission. Without seeing Europe it was impossible justly to estimate the assertion of travellers relative to the arts and agriculture of that interesting country, to distinguish the truths from the falsehoods contained in the infinity of books that treat of those subjects, or to adapt their precepts to the soil, climate, and habits of the United States. And it is no small source of happiness to me to believe, that however my public service may be appreciated, my pursuits in what may be thought a more humble line, are not altogether useless.

Having urged my fellow citizens to give some attention to the fine arts, and pointed out the easiest means of doing it, I see, with pleasure, other populous cities in my native country following the example set them (upon my suggestion) by New-York, in the establishment of academies for the fine arts. My ambition, however, leads me to render myself more extensively useful, by suggesting and enforcing such improvements in agriculture as may add to the wealth of individuals, and, by forming the basis of manufactures, to the independence of our country. My occupations in Paris kept me from collecting all the information on that subject which I could have wished; yet some things I have noted in the countries I have hastily visited, which, I trust, may furnish useful hints, and lead to useful experiments. Among other objects, my attention was forcibly attracted to one that at present occupies not only the agriculturalists, but the statesmen of Europe. It was long thought that the Merino sheep could only be raised advantageously in Spain, and that their migration was necessary to the perfection of their wool. Under the influence of this opinion, the rest of Europe submitted to be tributaries of Spain for this precious commodity; and so slow is the progress of agricultural improvements, that though an enlightened Swedish nobleman naturalized them eighty years ago, in a country little congenial to their native habits, yet it was long before his successful experiments excited public attention. France, after some abortive attempts, succeeded so fully as to open the eyes of the neighbouring nations. I saw and admired her beautiful flocks; and the inquiries I had the means of making of intelligent men from different parts of Europe, convinced me, that instead of degenerating, they had improved in every region to which they had been transported. Knowing the United States to be peculiarly adapted to short woolled sheep, I was eager to put them in possession of this invaluable stock. And I shall not envy the glory of the Argonauts (which probably consisted in bringing the fine woolled Mingrelian sheep into Greece) if I can successfully plant the Merinoes of Spain in my native land.

It unfortunately so happened, that during the greater part of my mission, a number of my fellow citizens were suitors at Paris for debts unjustly withheld; for relief from embarrassments into which the perplexed and ever-varying laws of trade, and in too many instances their own imprudence, involved them. As few days past in which I had not either justice or favours to ask for others, I thought it improper to ask the latter for myself, but hoped to attain my object (more gradually indeed), by selecting two pair of the finest Merinoes I could find, and sending them over under the care of one of my own servants; believing that so small a shipment would not be noticed, and intending to follow them by others. They arrived in safety in the spring of 1802, and were, I believe, the first couples ever imported into the United States. I became less anxious on the subject, because I had the satisfaction to learn that Col. Humphreys had succeeded, some time afterwards, in introducing a much greater number, direct from Spain, so that I believed a foundation was laid for their perfect establishment. After my return from Italy, being no longer in office, I obtained permission to ship others that Mr. Chaptal allowed me to select out of the highest bred flock in France. A variety of circumstances have hitherto prevented their arrival; but I still have the hope of seeing them here, with their increase since I purchased.

I was astonished when I found upon my return, in 1805, that the introduction of Merino sheep had excited little attention; and that although the Legislature of Connecticut had very properly noticed the patriotic exertions of Col. Humphreys, none of his sheep had been sold in this State. I had also the mortification to find, that notwithstanding my injunctions, mine had been much less extended than I expected. Nay, I learned with surprise, that a flock which consisted of near one hundred of one-half and three-fourths breed Merinoes, from a ram sent over by M. Delessert to his farm at Rosendale, near Kingston, had been sold at vendue at a price inferior to that of common sheep, and that above one half of them had perished from neglect the following winter. Such is commonly the case when novelties are introduced in agriculture, till the mind of the husbandman is prepared for their reception. I knew the importance of the object, and I resolved to leave no means unessayed to convince my fellow citizens of it. I began by purchasing all I could find of the scattered remnant of M. Delessert's flock. I picked up twenty-four ewes, and the price I paid for them attracted the notice of those who had seen and neglected them.

In 1806 I submitted to the Society of Useful Arts two essays on the subject of Merino sheep. They were received with a degree of attention which exceeded my hopes. The enlightened farmers were awakened to the subject, and the Legislature stepped forward and seconded their ardour by judicious encouragements.

Numbers of my fellow citizens are now endeavouring to supply themselves with this invaluable stock; and many who had never given the least attention to sheep, are extending their care to Merino flocks.

Finding myself frequently called upon for information, and being anxious to communicate all that my experience or inquiries had taught me upon the subject, as well as to keep alive the interest that I had happily excited in my fellow citizens, I believed that both might be effected by the publication of a little volume which should in some sort combine information with amusement, and, taken in connection with what I had before written, serve as a kind of Shepherd's Manual, and point out to the rich and the poor farmer the easiest means of converting their flocks into Merinoes, as well as the advantage that would accrue both to themselves and their country by the change. I have endeavoured, in the execution of this work, to render the style as simple as the subject of which it treats; to sketch the natural history of sheep in that rapid manner which would serve to satisfy a plain farmer, without swelling the work with disquisitions adapted to the taste of the experienced naturalist. I am extremely flattered by the attention the Legislature and the Society of Useful Arts have shown it, in deeming it sufficiently important to be printed at the public expense. Should it contribute to the extension of the Merino sheep, to the mutual advantage of the agriculturalist and the manufacturer, it will be very consoling to me to believe, after having devoted the prime of my life to promote the political interest of my country, that its decline is not absolutely useless; and that those whose fathers have shared in the *labours* of my youth, will receive some advantage from the *amusements* of my age.

ESSAY ON SHEEP.

THERE are few studies more generally amusing than those which relate to natural history, or rather to that branch of it which comprises the history of animals; it is sufficiently simple to be embraced by the untutored mind, and yet so comprehensive as to employ the faculties of the most elevated. The first will be entertained by the more obvious characters of the animal he considers, by its innocence or its ferocity, by its manners, its habits, and the instincts which lead it to provide for its wants and those of its offspring. A more profound philosopher will carry his views further; he will analyze the reason or the instincts of the animal, will examine the internal structure, and will admire the wonderful harmony that exists in the several parts of his body, and the analogy that is found between these and its manner of life. He will be insensibly led from the examination of the creature to a contemplation of the Creator, and will acknowledge his wisdom, and his goodness, in having exactly adapted the corporeal and mental faculties of every animal to the station he has been pleased to assign it in the scale of beings.

While the pride of man is humbled by the reflection, that the most profound works of art are but feeble imitations of nature, he will derive some consolation from the consideration, that God has condescended in some sort to render him his agent, and to give him extensive powers over the animal and vegetable creation; not only in subjecting them to his control, but even in enabling him, within certain limits, to change and alter their natures, so as better to adapt them to his own use, without subjecting them too far to his whims. The various species of grain and fruit that make his food. are no where to be found wild, but have been brought to their present state of perfection by the care and cultivation of man.

The flowers that bloom in the desert (with very few exceptions) are small and pale, for the most part single, and but slightly fragrant: to the culture of man they owe their brilliant and varied colours, their rich profusion of petals, and their high and grateful fragrance. Domestic birds and beasts change and vary their colours either to gratify his fancy, or to afford him natural marks by which to designate his property; while, in their native state, they wear an unvaried uniform, with now and then such an exception to this rule, as to afford a hint to man, and the means of grafting a permanent change upon accidental varieties.

The power of man to effect useful alterations in the animal creation, is in nothing more obvious than in those which sheep have undergone. It is impossible to see this animal overloaded with wool, slow in its movements, and possessed of no means of defence against its numerous enemies, without being convinced that such an animal could never exist in a state of nature. That it must therefore owe its imperfections to man, as it pays him by those very imperfections for his support and protection.

I have thought that the natural history of this animal, with an account of its varieties, would not be uninteresting to my fellow citizens; more particularly if it was accompanied by such didactic remarks as would contribute to the improvement and perfection of the breed in such manner as to enable us to draw the greatest profit from it.

It will easily be conceived, that little new can be offered relative to a quadruped that has so long lived under the care and observation of man; but it will certainly be useful to bring together the observations that have been made by different men at different periods, and to comprise in one little volume what must now be sought in various and expensive collections, written in different languages, and for the most part out of the reach of those of my fellow citizens to whom this is addressed.

I have already observed, that an animal which propagates slowly, which has no means of defence, and which invites by its extreme timidity the attack of its enemies, without possessing the agility to avoid them, could never have existed under its present form in a savage state, but must at all times have owed its protection to man. Should any country in which sheep exist be depopulated, the total extinction of the race would follow the depopulation: we must then seek for the original stock, or prototype of sheep, in some quadruped which possesses force, address, or agility enough to enable it to exist without the aid of man. Some have sought this in the goat; but this is evidently a distinct animal, though very nearly related; since the he goat will produce with the ewe a lamb without wool, that will be productive. But the ram will not impregnate a she goat, which marks an obvious distinction in the race, and shows that the sheep is more degenerate than the goat. Besides that, the goat is evidently descended from the Bouquetin, which appears to me to resemble the tame goat so strongly, that I have not been able to remark any difference in their looks, in their habits, or in their musky smell; except that the Bouquetin* is a larger and stronger animal than any species of domestic goat that I have seen. The horns also form a characteristic difference between the sheep and the goat. Buffon, and all naturalists since him, have supposed the Mouflon Musmon, or what is sometimes called the Argali, and which Linnæus distinguishes by the name of Ammon,† to form the stock from which the different varieties of domestic sheep have originated. Indeed, the resemblance of this animal to the sheep is so striking, that the Russians call it by the name of the Wild Ram. But it resembles the sheep as the vigour of manhood resembles the feebleness of infancy, or the decrepitude of age. The one possesses force, strength, activity; it can defend itself against the weaker animals, and elude the pursuit of the strong; while the other can neither fight or fly; but, without other defence than

* Hircus Sylvestris aut Ibex.

† Ovis Ammon.

its innocence, would soon be destroyed by that numerous host to which this is the feeblest of arms, if the utility of the race had not constituted man at once its tyrant and protector.

Pliny says that in the island of Corsica there is a species of Musmones not unlike sheep, whose covering is more like the shag of goats than the wool of sheep; and that the product of this animal with the common sheep was anciently called Umbri. From this circumstance it may be inferred, not only that they were occasionally mixed, but that the mixed race were so common as to merit a distinct name. This animal is not, however, confined to the island of Corsica; it is at this day to be found in all the uncultivated parts of the islands in the Archipelago, in Greece, in Sardinia, and in the north-eastern parts of Europe and Asia, even in Kamschatka and Siberia.* The following

* Pennant seems to think that it is also found in America, and, in proof of it, he says he has received from thence a fine fringe of twisted wool, which had ornamented the dress of an inhabitant of Red Jack, presented by Dr. Pallas, and that he had himself received another from the habit of an American of latitude 50. The first was white, and of unparalleled fineness; the other as fine, but a pale brown. The first he supposed the wool which grows intermixed with hair on the Argali, and the other to have been from the coat of the Musk Bull, which is a native of our country, and covered with extremely fine long hair, and beneath that a coat of very fine wool. The domestication of this animal would merit legislative attention. The missionaries to California in 1697, describe two distinct animals, with a head like a deer, and the horns of a ram, which they say were furnished with very good wool, and which they called Wild Sheep. These were doubtless the Musmones or Argali.

passage, translated from Professor Pallas's voyages, will serve as a full description of the animal. After noticing a summons that he had received from his troop of huntsmen who had killed a wild sheep and lamb, he describes the first in the following words: "The wild sheep called Argali by the Monguls, is stronger than a fallow deer, and weighs about 20 poud (or 660 lb). The ram weighs more, because his horns, when full grown, weigh sometimes more than a poud (33 lb); he is higher upon his legs than a tame sheep, and also more massy. I could remark but little difference in the formation of the head. The Argali has small upright ears. The horns of the female are of a middling size, and form crescents; they are also flat, with two blunt angles over the back, but the lower part forms a sharp angle in front. The horns of the male become enormous, and form a spiral on each side of the head, as those of the European ram; the tail is short, and the hoof like that of the common sheep; in winter the hair is long and frizzled, and mixed with much wool; on the contrary, it is short and smooth in summer. The old sheep had already (22d July) lost their winter coat, at least very little of it remained; their colour is an ash grey. This animal keeps upon mountains that are dry, desert, and free from wood, and upon rocks on which he finds acrid and bitter plants. The ewes lamb before the snows are entirely melted. The lambs resemble young roebucks. Their horns appear on their birth; their hair is soft, woolly, frizzled, and of a deep grey. The stag is not so wild as the Argali; it is almost impossible to approach him; when pursued he makes many turns to the right and to the left, and it often happens, when he finds no rocks or eminences to hide in, he turns upon his steps and passes before the face of his pursuers. He is astonishingly light and swift in the course, and can support a long pursuit. But, however wild this sheep may be, in its infancy the lambs are easily tamed, and habituated to drink milk and eat hay. The soldiers employed on the out posts have frequently ascertained this by experiments."

It is observable, that though there are strong marks of difference between the Mouflon and the domestic sheep, yet there are also strong points of resemblance. The first has been diminished by cultivation as inconvenient, while the last has been improved on account of its utility. As this quadruped has probably been found throughout all the mountainous parts of

Europe and Asia, and perhaps even in Africa; as its young are easily tamed; as its milk, its flesh, and its skin are extremely valuable to man in a savage state, it is highly probable that it was among the first quadrupeds that were domesticated; and, from this circumstance, it has perhaps wrought no less change in man, than man has in it. What respect do we not owe it, if, as is highly probable, we are indebted to it for the conversion of man from the wild and wandering savage, to the mild and gentle shepherd! The horse, the bull, and the camel, were probably conquests subsequently made over the animal creation, because it required more strength and skill to tame and render them useful; but the young Mouflon was soon tamed; the female savage that followed her husband to the chace snatched it from its bleeding dam, pressed it to her bosom, and became its mother; it sported with her children, and taught them to love a race which they had hitherto pursued only to destroy. A slight ray of reason must have shown the savage how much less precarious his subsistence would be, if he could draw it from an animal that fed at the door of his hut, than if he was compelled to seek it in the chace. He would extend his flock; he would cease to trespass upon the hunting grounds of others; but he would appropriate a portion for the support of his flock; he would compound with his tribe; or the whole tribe, going into the same culture, would mark out limits which they would not suffer to be trespassed upon; they would unite for common defence; the rights of property would be known, and a nation be formed where before only wandering hordes had existed. By what simple means does providence produce the greatest good? That we are not at this moment fierce, savage, and brutal, little superior to the beasts that roam in the wilderness, and only employing that little superiority in their destruction, and in the destruction of each other, is probably owing to the domestication of graminivorous animals, and, first of all, to that of sheep. To them we are also indebted for some of the most pleasing, as well as for the most important and useful arts. The cradle of music and poetry was rocked by the shepherds of Arcadia; while the spindle and the distaff, the wheel and the loom, originated in the domestication of sheep. This little animal then, in losing its own wild nature, has not only converted the savage into the man, but has led him from one state of civilization to another; the fierce hunter it has changed into the mild shepherd, and the untutored shepherd into the more polished manufacturer. The more sedentary men became, the greater were their wants and dependence upon each other; and in those wants and that dependence originated civilization and polished societies.

The sheep which approaches nearest to the original stock, and has suffered less by the art of man, is the Adiman or African sheep. These are large, active, and covered with hair without any intermixture of wool. It is observable, that the Mouflon or Argali has a fleece composed of hair and wool in the cold climate of Tartary; yet those in warmer climates have no wool. It is probably from this stock that the sheep of Guinea have been reared; and as they belonged to a people to whom woollen clothing would be of no use, who formerly went naked, and if they now wear a slight and partial covering, it consists of lighter materials than wool would afford; it is, therefore, not surprising that they have not added to the degeneracy of their flocks by rendering them woolbearers. Half savage themselves, they are content that their domestic animals should resemble them; since in that state they are fitter to furnish food than if part of their sustenance went also to the supply of clothing. It is generally supposed that the want of wool is the natural effect of the climate; and that the wool-bearing sheep, upon being transported to low latitudes, loose their wool, and acquire hair; and the smooth skinned sheep that are found in most of the West-India islands are adduced as a proof of this theory.

I will not pretend to say that climate, in a long series of years, may not produce a change in the nature of quadrupeds; but if it does, I believe it must operate very slowly, and much more gradually than is generally supposed. The hairy sheep* that are found in most of the islands appear to me to bear evident marks of African origin; like those, the rams and ewes have a kind of dewlap of long hair pendant from their necks; they are larger and more active than the common European sheep. It is certainly not to be wondered at, that countries which maintain a constant intercourse with Guinea should have brought over their sheep as well as their men, and, as this breed are better adapted to a warm climate than the sheep of Europe, they have probably become the predominent sheep in the islands; though there are, in many of them, wool-bearing sheep, which remain unaltered, except by mixture of

* Ovis Aries Guiniencis.

their blood with that of the Guinea sheep. I have myself had occasion to make some experiments in those sheep, three of which were sent me by my worthy friend Mr. Kerby, from the island of Antigua. I had these several years without observing that any sensible change was produced by exposure to the air during our cold winters, except that, like the Argali, they acquired, the first winter, a coat of very short and fine wool below their hair, which fell off, together with the hair, as the summer heats came on, when they acquired a new coat of hair only; and, as winter approached, this was again thickened by an under stratum of very fine short wool. These sheep are then the Argali, but moderately degenerated. Indeed, it would appear a little extraordinary, if the climate that converts the hair of man into wool, should, by some retrograde movement, change the wool of sheep into hair. The next in order, in point of degeneracy from the original stock, is the sheep of Iceland.* Like the Argali, they have two coats; one of extremely coarse hair, which hardly merits the name of wool, and another beneath it of a softer and finer quality, but so mixed as to make it impossible

* Ovis Aries Polycerata. *Lin.*

to separate them. These sheep partake of the hardiness of the parent stock. The large horns peculiar to the Argali is not, indeed, found among them in the same form, but it is broken down into several smaller branches. Most of them carry four, and many five horns of considerable size, and always spiral. What is remarkable, and shows that this circumstance is owing to the address of man, and not to the effects of climate, is, that when the common sheep are brought to Iceland, their horns diminish or disappear altogether; this, at least, is affirmed by V. Bomari. The Iceland flocks are never stabled, but seek their food by following the horses and eating the grass and moss that they uncover; their own feet being too feeble to dig the snow. Their shelter is the jutting rocks, or mountain's caverns. At the approach of a snow storm they run violently towards the sea, and are sometimes precipitated into it by each other. They have probably learned from experience that the sea softens the rigour of the air, and that the snow is sooner dissolved in its vicinity than upon the mountains. If they are surprized by a snow storm before they can reach the coast, they turn their heads toward each other, and patiently expect, under their fleecy covering, the aid of their owners, who do not fail to search for and relieve them as soon as possible. They distinguish the spot in which they are buried by an exhalation which arises from their breath. If this aid is so long delayed as to subject the sheep to the danger of starving, they reciprocally feed upon each others fleece. This race is extended through the Danish islands; where it is equally neglected during the winter; and their instincts improve by this neglect. They keep each other warm by pressing close together when the bleak winds pinch them; and those from the centre relieve in turn, those who, in the outer part of the circle, are exposed to the severity of the blast: thus necessity sharpens the invention of beasts as well as of men. Left to themselves, and compelled to rely upon their own resources, they know how to call them forth; while our helpless sheep, who rely wholly upon the attention of their keepers, will frequently suffer from cold, rain, and snow, without moving into the shelter that is provided for them.

I find in the Supplement to Buffon the drawing of another species of sheep, which he calls the Wallachian Sheep, but it is accompanied by no description. It is the sheep called the Strepsiceros or Cretan Sheep, and only differs from the common sheep in having horns spiral, and growing upright. This sheep has in Wallachia and Hungary an undulated wool, which is valuable for peltries; but I imagine this is rather the effect of art than nature, for I find that the mongul Tartars make use of the following means to have their peltries of this sort. The lambs have naturally with them, as they frequently have with us, a kind of wave or curl in their wool when they are first dropped. In order to improve this, and to render it permanent, they cover the lamb with a linen coat, tied close about the body. This they water frequently with warm water, and loose it occasionally as the lamb increases in size. When it has attained the necessary perfection, of which they judge by inspection, they kill the lamb. These skins are more valuable than any of the furs, except those of the Sable. It might be worth the trial here upon some of our lambs whose wool is most curled and waved when they are dropped. By this means a new source of profit might be derived from this useful animal; nor would the flesh be lost: a lamb is fit for the table at a month old. I have seen hundreds of them sold by butchers at Naples much younger. There, as in Spain, where they have migrating flocks, probably half the lambs are killed. In

Spain every lamb of the migrating flock has, besides his natural, a foster mother. In order to induce the last to take the lamb, the skin of that which has been killed is put upon the one to be raised, and, in this disguise, she mistakes it for her own, and gets familiarized to it in a few days. This does not, indeed, always succeed. When it does not, the shepherd compels her to admit the lamb by tying her.

The race of sheep that I shall next notice, is one that is more extensively diffused than any other, since it is found throughout Asia, and a great part of Africa, as well as through the north-eastern parts of Europe. I refer to the broad-tailed sheep.* These differ, as the ordinary European race, in the nature of their covering. In Madagascar, and some other hot climates, they are hairy; at the Cape of Good-Hope they are covered with coarse harsh wool; in the Levant their wool is extremely fine, or, in other words, they are adapted to the necessities of the people by whom they have been changed from their wild to their domestic state. These sheep are generally larger than those of Europe, in which circumstance only, and the form and size of their tails, they differ from

* Ovis Aries Laticandata.

them. The broad-tailed sheep are of three species. In the one the tail is not only broad but long, and so weighty that the shepherds are compelled to place two little wheels under it to enable the sheep to drag it. These tails are said sometimes to weigh from forty to fifty pounds. Another species have the tail broad and flat, but not very long, covered with wool above, but smooth below, and divided by a furrow into two lobes of flesh; these are also said to weigh above thirty pounds: I should not however, estimate the weight of those which I saw in the menagery at Paris, at more than ten or twelve pounds. In some species a small thin tail projects from the centre of this fleshy excrescence. The composition of this excrescence is said to be a mixture of flesh with a great proportion of fat, and to be very delicate food; but the animal has little other fat, the tail being in him the repository of that fat which lays about the loins of other sheep. In cold climates the fat of the tail resembles suet; but in warm ones, as at the Cape of Good-Hope, Madagascar, &c. it is so soft, that when melted it will not harden again. The inhabitants mix it with tallow in certain proportions, when it assumes the consistency of hog's lard, and is then eaten like butter, or used for culinary purposes. Naturalists imagine that this excrescence is owing to some circumstance in the food of the sheep, which makes the fat fall down from the loin into the tail, and thus occasions this monstrosity. I do not, however, think this probable, since the prodigious extent of country through which this race is propagated, must render the food as various as the climates in which they are bred. I rather think, that it owes its origin to the art of man, grounded upon some of those sports of nature, which, in all domestic animals, afford a basis whereon to ingraft his whims. The broad-tailed sheep does not differ more from the Argali, than the white fan-tailed pigeon does from the wild blue European pigeon from which it originally descended; or than the little hairless smooth-skinned Turkish cur, from the rough shepherd's dog, the common ancestor of his race. It may be asked, to what end would man cultivate this deformity, and that too through so extensive a region as to forbid our attributing it to whim or fashion? May not the shepherd who first observed this lusus naturæ in his flock have concluded, that he had made a very valuable acquisition, since he not only had a sheep that gave him as much wool, milk or flesh as the rest of his flock, but a tail, which, in addition, gave him a comfortable meal, or what is still more valuable among savages, plenty of grease for his toilet and his kitchen? This circumstance alone would make him attentive to cherish and propagate the deformity; and the rather, as he must soon have found that it was attended with another important advantage; the sheep being more unwieldly, would be less apt to stray or return to its savage state; an object of considerable importance in the early state of society. We find at this moment a deformity in sheep cultivated with attention among ourselves. An accidental variety of sheep have been found here with short crooked legs, such, in fact, as to cripple them, and to make motion, as I should think, painful to them. These, called the Otter Sheep, are valued for this deformity, because it disables them from straying or leaping over walls or fences; and what was at first probably an accidental circumstance, has become the basis of a new and unsightly race. If a civilized nation, with whom taste has formed a standard for beauty, can consent to cripple God's works, and erect an altar to deformity, whereon to sacrifice the enjoyments of a helpless and useful animal, why should we be surprized, that savages, ignorant of the beauty of proportion and the harmony of forms, should have early sought to curb the troublesome agility of their sheep, by giving the same preference to rickety tails, that some among us have done to rickety legs?

I come now to speak of those breeds of sheep that are best known to us, and indeed the most useful in our state of society—the sheep of Europe. I should however first observe, that some provinces of Persia possess a breed of sheep whose wool is finer and more valued than that of Spain; but as I have no where met with a minute account of them, I shall proceed to notice the race of sheep which holds the first rank, and bears the finest fleeces of any known in Europe—I mean the Merino sheep of Spain. The race varies greatly in size and beauty in different parts of Spain. It is commonly rather smaller than the middle sized sheep of America. The body is compact, the legs short, the head long, the forehead arched. The ram generally (but not invariably) carries very large spiral horns, has a fine eye and a bold step. The ewes have generally no horns. The wool of these sheep is so much finer and softer than the common wool, as to bear no sort of comparison with it; it is twisted and drawn together like a cork-screw;

its length is generally about three inches, but when drawn out it will stretch to nearly double that length. Though the wool is, when cleaned, extremely white, yet on the sheep it appears of a yellowish or dirty brown colour, owing to the closeness of the coat, and the condensation of the perspiration on the extremities of the fleece. The wool commonly covers great part of the head, and descends to the hoof of the hind feet, particularly in young sheep; it is also much more greasy than the wool of other sheep. Spain contains besides the Merinoes, a variety of other sheep. Those called the Choaroes are much longer, larger, and higher upon the legs than the Merinoes. Their heads are smaller, and deprived of wool. Their constitutions are stronger than those of the Merino. Their wool near eight inches long, but straighter and of less value than that of the Merino. This race extends through all Spain, even into those provinces in which the Merino is most perfect. The other sheep are a mixed breed between those and the Merino. The number of these two species is computed at about 6,000,000. Among the Merinoes there are varieties, probably occasioned by the care or fancy of the original cultivators of this valuable stock in different parts of Spain. Castile and Leon has the largest, with the finest coats. Those of Soria are small, with very fine wool. Those also of Valencia, which, like the last, do not travel, have fine wool, but of a very short staple. The greater part of the Merinoes of Spain are transhumante, and migrate from the south to the north, and from the north to the south twice every year. This has probably contributed to the health of the sheep, and, as a consequence of it, to preserve the beauty of the wool, without however being essential to it; as appears from the fine wool produced by the stationary flocks that I have mentioned, and other stationary flocks in the hands of individuals, whose wool is not inferior to that of the migrating sheep. Spain is bounded to the north by mountains of such altitude as to be covered during the winter with snow. These however afford fine pasturage in the spring and summer, when the plains in the south are parched by the sun. It was very natural then for the shepherd to avail himself of this circumstance, and while the country was little cultivated, to drive his flock from the burnt grass of the plains to the fresh and verdant herbage of the mountains; and again, when this was chilled by frost, or covered by snow, to return to the plains that had regained their verdure; the winters of

Spain not being so severe as to destroy the vegetation, except in the mountains. Necessity also contributed to keep up this practice. During the long wars that were carried on between the natives of Spain and the Moors, agriculture was neglected, and the only property which could be saved from the ravages of an enemy, was that which could be easily removed; but they were content at that time to travel only from the plains to the adjoining mountains, and not as at present to travel the whole kingdom twice a year. Neither convenience nor necessity can be offered as an excuse for a practice so hurtful to agriculture. This was founded in abuse of power. Happy would it be for mankind if this was the only instance in which tyranny and oppression had been engrafted upon necessity.

The greater part of the travelling flocks in process of time got into the hands of the sovereign, or into those of the principal courtiers and clergy; and from thence we must probably date the oppressive code by which their march is regulated, and the origin of the great Council of the Royal Troop (Consejo de la Mesta), by whom those laws are administered. M. Lasterie, in his excellent treatise, gives the following account of this council.

"The Mesta, which originated with the times, in which force only gave law, about the middle of the fifteenth century, formed a political body. This association was composed of rich and powerful persons, and some monks, all proprietors of flocks, which, under the authority of government, made laws and decided questions relative to pasturage and flocks of sheep. Two great quarto volumes formed the code of privileges, and the arsenal in which were found arms to combat justice and oppress the weak. It was seldom that proprietors of land made demands when they sustained damage, thinking it better to suffer than to contest, when they were assured that the expense would greatly exceed any compensation they might recover. It is sufficient to say, that this tribunal is not only adverse to the enclosing of land, but that, under some circumstances, it may prohibit proprietors from cultivating their inheritance. A Spanish writer (Jovellanes), in a memoir addressed to the King of Spain, says, 'the corps of Junadines (the proprietors of flocks) enjoy an enormous power, and have, by the force of sophisms and intrigues, not only engrossed all the pastures of the kingdom, but have made the cultivators abandon their most fertile lands: thus they have banished the stationary flocks (the estantes), ruined agriculture, and depopulated the country.' It is easily conceived, that five millions of sheep traversing the kingdom in almost its whole extent, for whom the cultivators are compelled to leave a road through their vineyards and best cultivated lands of not less than ninety yards wide, and for whom, besides, large commons must be left; I say, it is easily conceived that such a flock must greatly contribute to the depopulation of the country, and that the revenue that the King draws by the duty on wool is snatched from the bread of his people."

When the severe weather commences upon the mountains, the shepherds prepare to depart, which is generally about the end of September and throughout the month of October, to seek more temperate climates and fresher pastures. In April and May, according as the season is late or early, they return to the mountains. They generally travel about five or six leagues a day, and stop occasionally in the pastures prepared for them: the head shepherd precedes, and the rest flank or follow the flock to collect the stragglers. Like Virgil's Libean shepherds, they carry every thing with them.

Omnia secum
Armentarius Afer agit, tectumque, laremque,
Armaque, Amyclæumque eanem, Cressamque pharetram.

This is comprised in a very short catalogue. The skins of sheep that serve for their beds, a kettle, a leather bottle, a knapsack, a spoon, a lancet to bleed their sheep, a scissors, a hatchet, a knife, and bread and oil or suet, on which they subsist, and a few drugs for their sheep. These, with the skins of those sheep that die in the passage, are carried by a few beasts of burthen which accompany the flock. To facilitate the march, a number of wethers of the largest size, which they call Mansos, are rendered very tame. These carry bells, and are taught to obey the signals of the shepherds, and either march or stop as they direct. The rest of the flocks follow their leaders. As soon as they arrive at their winter quarters, the shepherd's first care is to form the pens in which they are gathered at night to protect them from the wolves, who always migrate with the sheep, in order to pick up the sick, the weak, or the stragglers. These folds are made of genista hispanica, which is a soft rushy shrub: mats, baskets and ropes are made of it. The meshes of these net enclosures are a foot wide. The dogs, which are of a large breed, serve to guard this fold at night. The shepherds make their own tents with stakes, branches, and brambles, and have for this purpose a right to take one branch from every forest tree. Ten thousand sheep compose a flock under the direction of one chief, and this is divided into ten tribes. The head shepherd has absolute dominion over fifty shepherds, and as many dogs, five of each being annexed to a tribe. His salary is about two hundred dollars a year, while that of the first shepherd of a tribe is only ten, the second eight, the third and fourth still less, and a boy only two and a half. Their daily allowance of food is two pounds of bread, and as much to each dog. They may keep a few goats or sheep, of which they have the meat. but not the wool. They receive as a gratuity about six shillings in April, and as much in October, by way of regale. On the road they are every day, at all seasons, exposed to the air, and at night have no shelter but their miserable huts. In this way live to a considerable age the twenty-five thousand men that compose the shepherds in Spain. The flocks consist of rams, ewes, wethers and lambs, in the following proportion: five rams, one hundred ewes, twenty-five wethers and fifty lambs. The small number of lambs is owing to the shepherds killing all that are not necessary to keep up their stock, which is, of course, limitted by the right of pasturage. The number of travelling Merino sheep is about five millions. The fleeces of the rams weigh eight and a half pounds, of the ewes five, which loses half in washing; but in this there is great variety, according to the different species of Merinoes. The produce is about twenty-four reals, or sixteen shillings per head. Of this the owner receives but two, the King six, and the remainder goes to the payment of expenses, of pasture, tythes, shepherds, dogs, &c. When the sheep return to their summer pasture they have as much salt given them as they will eat. One thousand sheep are allowed one hundred pounds of salt, which they eat in about five months. They eat none when on their journey, or in their winter quarters. They suppose in Spain that salt contributes greatly to the fineness of the wool. The shepherd places fifty or sixty flat stones at about five paces from each other; he strews salt upon, and leads the sheep among them. In the month of April it requires some vigilance to prevent the sheep from marching off without their shepherds, to the very place where they fed the preceding year, which they sometimes do to the number of three or four hundred in a flock.

As the Merino sheep are greatly superior to any other in Europe, it has naturally led to an inquiry into their origin, and the time of their introduction into Spain. On this subject history does not afford all the light we could wish. Many suppose that they were originally introduced from the coast of Barbary, by Don Pedro the fourth, who ascended the throne of Castile, in the middle of the fourteenth century. Others again attribute their introduction to Cardinal Ximenes, who became Prime Minister of Spain in the beginning of the sixteenth century. And Anderson insists, upon the authority of Stow and some old records, that they were introduced from England as early as Edward the fourth, who died in 1483. Though all these circumstances may have contributed to improve certain breeds already existing in Spain, yet it is certain that the fine woolled sheep were found in that country at a much earlier period. Strabo, speaking of the beautiful woollen clothes that were worn by the Romans, says, that the wool was brought from Truditania in Spain. After the conquest of that country by the Romans, colonies were planted there, who carried with them the arts and love of agriculture which distinguished that nation of warriors.

Columella (uncle of Columella who has left us an excellent treatise on agriculture), a rich colonist, who lived at Cadiz during the reign of Claudius, and made agriculture his pleasure and his pursuit, was struck with the beauty of the wild rams that were brought from Africa to be exhibited at the Roman games. He coupled those with Tarentian ewes, which were celebrated for the softness of their wool, and procured by this means a race whose fleeces resembled that of their dam in softness, and that of their sire in the colour and fineness of the wool. Whether any permanent change was effected by this experiment of Columella's, I know not; but as Spain was at that time highly civilized, and as agriculture was a favourite pursuit of all who were not occupied in war, I think it highly probable that this experiment laid the foundation for a general improvement in the sheep of the country. If it did, Spain is more indebted to the patriotic efforts of one enlightened farmer, than to the ablest of her statesmen. How much should it excite the laudable ambition of virtuous men to know that there is no condition in life in which they may not be useful, and that God has often made a simple farmer, or a plain mechanic, the means of diffusing his blessings upon mankind. Many centuries elapsed after this in which we are left in the dark as to the history of agriculture in Spain. The conquest of the country by the Goths, and the subsequent reduction of it by the Moors, together with the long wars between the latter and the native Spaniards, have cast a veil over their history; but as the Moors were industrious agriculturalists, and kept up their connection with Africa till their first conquest by the Spaniards, it may be presumed that they pursued the path marked out by Columella. It is probable, however, that during that disastrous period which preceded their expulsion, the farther improvement of this useful race of sheep had been neglected; and as in human affairs scarce any thing is stationary, it is also probable that they were suffered to degenerate: for we find, as I have said, Pedro the fourth, more than thirteen hundred years after the death of Columella, reviving his experiments on an enlarged scale, and introducing a great number of sheep from Barbary. His efforts were crowned with success, and Spain became in the fourteenth century what she had been in the time of the Romans, famous for the fineness of her wool. The race was again renewed from Africa by Cardinal Ximeníes, two hundred years afterwards. From these circumstances it is highly probable that Spain owes her Merino race to the mixture of her native sheep with those of Barbary, though (as often happens) neither, in their native state, may be equal to that produced from the union of both. This may account for the sheep of Spain being at present superior to those of Barbary, though in part descended from them.

The wool of the Barbary sheep is glossy and fine (at least such as I have seen of it), but wants the curl of Spanish wool. I may here mention a fact which in some sort supports this assertion, though an isolated fact ought not perhaps to serve as the foundation for a theory. I have in my flock a ewe that is descended from a Barbary ram. Her fleece is long, straight, and fine, and in every particular except the last unlike the Spanish wool. I have three lambs from her by a Merino ram; the wool of each of these is nearly equal in fineness, softness, and elasticity, to that of their sire, and would at least be taken for that of a seven-eight breed Merino.

I cannot think, with Mr. Anderson, that the fine wool of Spain is derived from the stock of England, though it may be admitted that British sheep have been imported into Spain, as it appears by custom-house entries that English wool was also exported to Spain, which was at that time a manufacturing country, and supplied England with cloth. For many of their manufactories the long wool of England might have been found useful, and it might also have been thought desirable to propagate the breed that bore it, without any intention of degrading the Merino breed. It is possible that the long-woolled sheep of Spain, which are called Choaroes, and are much larger than the Merino, are the descendants of the English sheep, mixed with the common sheep of the country. Had England possessed in the fifteenth century, the fine race which is now the pride of Spain, it is hardly possible to suppose that so shortly after as the reign of Henry the eighth, the breed should be so entirely lost as to induce that prince to import, by permission of Charles the fifth, three thousand Spanish sheep, and to disperse them through his kingdom, placing them under the care and superintendence of commissioners specially appointed for that purpose. In fact, it was not till the reign of his father that woollen cloths were manufactured in England to any extent, and none I believe were for many years after exported from thence.

Under these circumstances more attention would naturally be paid to the carcase, and to the quantity than to the quality of the wool. In size and weight of fleece the English sheep, generally speaking, exceed that of any other part of Europe.

Sicily also possesses a breed of fine-woolled sheep, which migrate like those of Spain, but are inferior to them in the quality of the wool. Those, with most of the sheep I have seen in Italy, have pendent ears. From this circumstance I presume they have been longer domesticated than those of Spain or other parts of Europe. And as this country was originally settled by the Grecians, it is highly probable that the present race is from the stock of the first colonists: for, extraordinary as it may appear, notwithstanding the various changes which that country has undergone, its agriculture seems at the present to be what the poets describe it to have been two thousand years ago; and the implements of husbandry dug up at Pompeia and Herculaneum are evidently the models of those now in use in the vicinity of Naples. I consider pendent ears as a proof of very ancient domesticity, because I believe all wild animals carry theirs erect; and most, if not all of them, have the power of moving them to the point from which the sound is derived. When they cease to be their own protectors, and rely upon man both for defence and support, the organs given them with a view to these objects, are gradually impaired, and the debility which results from their inaction changes their very form.

The sheep of France and Germany have nothing particularly worthy of notice, if we except the improvements made within a short period by the introduction of Spanish sheep, on which I shall have occasion to speak more at large hereafter. The common sheep of the country have in general coarse fleeces, and not very heavy ones. Those of Rousillon and Berry must, however, be excepted. The first is in some degree mixed with the Merino, and partake of their qualities; and the wool of Berry is generally estimated at about eighteen cents the pound, while that of the common flocks does not exceed seven cents. In many parts of the country their carcases are large and heavy; but that held in the highest estimation is from Brittany, which is extremely small, but the best flavoured mutton I have ever met with. In French Flanders they have a large race of long-woolled sheep. They are not very numerous, requiring richer pastures and better treatment than sheep generally receive in France. Before I quit France it may be proper to speak of the introduction into that country of the Merino sheep, and of their great improvement.

It having been fully ascertained, by a variety of experiments, patronized by the administration, and conducted by enlightened agriculturalists, that the Merino sheep might be acclimated in France, without any change in their wool, application was made by Lewis the sixteenth to the King of Spain, for permission to export from thence a number of Merinoes. This was not only granted, but orders were given by the Spanish Monarch that they should be selected from the finest flocks in Spain. In the year 1786 four hundred rams and ewes arrived in France, under the care of Spanish shepherds. These are said to have been so much superior to any that had before been introduced, as not to admit of any comparison between them, which will easily be credited by those who know the difference between picked sheep and a whole flock taken together, even when the sheep are of one race. But the Merinoes differ essentially from each other even in Spain; those of Castile unite size and beauty to fineness of wool; the Leonese and the Segovians equal them in the latter particular, but fall far short in the former; but the sheep of the Escurial are the finest in Spain. The difference between the Merinoes that compose the national flocks of France and those lately imported from Spain, under the treaty of Bale (though these also are picked sheep), is so striking, that we can hardly attribute it solely to the care and attention which they have received in France, though much is doubtless due to this circumstance. Fortunately for France, the improvement in sheep begun under Lewis the sixteenth was continued through a revolution, in which almost every other useful institution was involved in ruin. A committee of agriculture was formed in the Convention, and under their protection the amelioration of the Merino flocks happily progressed. From these flocks a number of rams and ewes are annually sold, after the finest are picked out to keep up the original stock. It is very conceivable that this attention must contribute greatly to the improvement of their stock. It is remarkable, that though in pursuance of an article in the treaty of Bale, five thousand Spanish sheep have been introduced by the government, and a great number by individuals, and for the term of twenty years rams and ewes have been annually sold from the national flocks, yet the price of rams drawn from those flocks is daily increasing. This fact shows in a very striking point of view the advantages of this breed of sheep, since they have been enabled to conquer the prejudices even of the French peasantry, who adopt improvements very slowly. Having mentioned the superiority in size and beauty of the national flocks of France, it may be satisfactory to know the quality of their wool. This I shall give from the report of M. Gilbert, one of the members, to the National Institute of France.

" The stock from which the flock of Ram-
" bouillet was derived, was composed of indi-
" viduals beautiful beyond any that had ever
" before been brought from Spain; but having
" been chosen from a great number of flocks, in
" different parts of the kingdom, they were dis-
" tinguished by very striking local differences,
" which formed a medly disagreeable to the
" eye, but immaterial as it effected their quali-
" ty: these characteristic differences have been
" melted into each other, by their successive al-
" liances, and from thence have resulted a race
" which perhaps resembles none of those which
" composed the primitive stock, but which cer-
" tainly does not yield in any circumstance to
" the most beautiful in point of size, form, and

"strength; or in the fineness, length, softness, "strength, and abundance of the fleece. The "manufacturers and dealers in wool, who came "in numbers to Rambouillet this year (1796), "to purchase, unanimously agreed to this fact, "at the very time that they were combining to "keep down the price. The comparison I "have made with the most scrupulous atten-"tion, between this wool and the highest priced "of that drawn from Spain, authorizes me to "declare that of Rambouillet superior; unless, "as they pretend, the best of the Spanish wool "is not imported into France, but reserved for "England and Holland; an assertion which is "certainly very improbable, and which, if true, "would argue a great superiority in our manu-"factories; since the superiority of our fine "cloths over those of any other nation has "never been contested. All the wool of Spain "that I have examined, not excepting the "prime Leonese, the most esteemed of any, "appeared to me to contain much more jar "than that of Rambouillet. Every thing seems "to evince that we shall soon totally banish "this hard intractable hair, so hurtful to the "manufacture, from our fleeces. Almost all "the fleeces of the rams of two years and up-"wards, weigh from twelve to thirteen pounds;

"but the mean weight, taking rams and ewes "together, has not quite attained to eight "pounds, after deducting the tags and the "wool of the belly, which are sold separately." It is proper to observe here, that the French pound is about one-twelfth heavier than the English; but at the same time to note, that, from the general custom of folding the sheep in France, of feeding them in fallows, and wintering them in houses, they are very dirty, and their fleeces of course proportionably heavier: the loss in washing is about 60 per cent. so that the average weight of the ram's fleece would be, when washed and scoured, about six American pounds, exclusive of tags and belly wool.

Before I quit Europe it may be proper to take a cursory view of the English sheep, since, next to Spain, no country in that quarter of the globe is so celebrated for its wool; nor is there any that have paid so much attention to the improvement of their stock; insomuch that Young, in his Annals of Agriculture, asserts, that Bakewell, in the year 1789, was in the receipt of three thousand guineas a year for the hire of rams, seven of which brought him two thousand guineas. A spirit like this, attended with proportionate wealth, could not fail, in any country, to effect the most important improvements. It would be tedious and unnecessary to enter into a minute enumeration of all the varieties produced by different crosses, and other accidental causes in a kingdom which contains such a variety of soil and climate as Great-Britain, and in which the farmers have endeavoured to conform the breed to their situation; and the rather, as I have already noticed many of them in a paper read to and published by the Society of Useful Arts. Anderson divides the native British sheep into three sorts; the Highland breed, or rather the breed of the Western Islands, those in the Highlands being so far adulterated as not to be found in their original purity. These sheep, though delicate in appearance, are small and hardy. The wool is distinguished by a silky gloss to the eye, and a peculiar softness to the touch. It is not frizzled like the Spanish, but rather longer, and gently waved. When compared with the best Spanish wool in the London market, it was found to be finer in the proportion of seven to five. Stockings have been made of it at Aberdeen that sold at five and six guineas a pair. The wool of this breed, however, is either naturally, or by adulteration, very much mixed with hair or jar, so as to render the separation very difficult. The second is the short-woolled sheep of England and Wales, that yield the clothing wool: of these there are very great varieties. Few however that I have met with yield better wool than the common sheep of our own country,* and in general their wool is much worse, with the exception of one or two races, whose fleeces are very short and light, and sell at about forty-eight cents the pound. The South Down is at present the favourite, next to the Leicestershire or Bakewell breed. The South Down, both for size, quantity, and quality of wool, very much resembles the best of our sheep in the hands of good farmers. Their fleeces weigh from three and a half to four pounds, and sell at thirty cents per pound. Neither of these breeds yield wool of sufficient fineness for broadcloths of the first, second, and third qualities; these are all made from Spanish wool of different degrees of fineness, without admixture. Of this wool near seven millions of pounds are annually imported into Britain. The third distinct breed of England, and which is peculiarly their own, is the sheep that carry long wool fit for combing; and in this race they excel, I believe, every

* I speak of the Northern, not being well acquainted with the Southern States.

other part of the world. The wool of some of this family is very coarse, and only fit for blankets and carpets, and sells in England at about nine cents the pound; but then the sheep are extremely large, and their fleeces proportionably so, averaging about twelve pounds the flock round, and some have been known to carry above twenty pounds. Young mentions a fleece of twenty-seven pounds. Others, and more valuable races of long-woolled sheep, bear a fine white silky fleece, from which the finest worsted and camblets are made. This race is very numerous, and their wool may be considered as the true staple of British wool. Upon this breed Bakewell has engrafted his celebrated stock, or the new Leicestershire breed. The principle upon which he formed his system was, that those animals were most valuable which carried their flesh upon the most valuable parts, and were at the same time maintained with the least food. Wool was not his object, and accordingly his sheep are of the long-woolled breed, with wool of moderate length and medium fineness, and sells for nine-pence sterling. Fat upon the rump and ribs he considers as more important than tallow, and accordingly he has produced sheep on which it is there formed five or six inches thick. His sheep are, however, on that account, less valuable to the epicure than to the labourer, with whom they in some sort supply the place of pork. He insists that they require less food than other sheep; yet, in a comparative trial made between them and a Merino ram by Young, it appeared that they eat more, and gained less weight than the Merino, in the proportion of three to two. Small bones, a straight back, and broad chine, with short legs, are the favourite points in this new breed; and, indeed, they contribue very much to improve the appearance of the animal, and should be sought in whatever breed we cultivate, if they can be reconciled with the other essential qualities that we seek in sheep. Of the advantage of short legs I have, however, great doubt in a country which abounds in snow. Some judgment may be formed of the nature of British flocks by the average prices of their native wool, which Gov. Pownal, in a letter to Arthur Young, states as follows: Coarse seven and a half-pence, common eight and a half-pence, fine eleven-pence the whole fleece; at that time they paid six shillings and six-pence sterling per pound for Spanish wool, and now pay seven shillings and three-pence.

It would almost have been unnecessary to

notice the American breeds of sheep, since those who will interest themselves sufficiently in the subject to read this essay, can hardly be unacquainted with the breeds of their native country, had it not been that three kinds of sheep, till lately unnoticed, have attracted the public attention; the Otter, the Arlington, and the Smith's Island sheep. The Otter sheep, it is said, were first discovered in some island on our eastern coast, where I cannot precisely say, and from thence they have spread to the adjoining states. The sheep of this breed are rather long-bodied than large, and will weigh like the other sheep of the country, about fifteen pounds a quarter when killed from grass. Their wool is of a medium fineness, and a medium length; it is neither properly short-clothing wool, nor is it of such length as to be advantageously combed. But what particularly characterizes these sheep, and from which, together with the length of their bodies, they probably took their name, is the extreme shortness of their legs, which are also turned out in such a manner as to render then rickety. They cannot run or jump, and even walk with some difficulty. They appear as if their legs had been broken, and set by an awkward surgeon. To me there is something so disgusting in the sight of a flock of these poor lame animals, that even a strong conviction of their superior utility could hardly induce me to keep them. The only advantage that can result from this deformity, is, that they cannot pass over stone walls, and are confined by slight fences. Whether this will counterbalance the sufferings to which they must be liable in a deep snow, the impossibility of driving them to distant pastures or to market, and the facility with which they may be destroyed by dogs, is a matter of calculation with economical farmers. Those, however, who possess a grain of taste, who take a pleasure in the sportive gambols of their lambs, and who delight rather in perfecting than in maiming the works of nature, will seldom be induced to propagate, beyond what is absolutely necessary, an infirmity which abridges the short enjoyments of a useful and helpless animal.

From these sheep I turn with pleasure to the Arlington long-woolled sheep. These Mr. Custis, who was the original breeder of them, informs me were derived from the stock of that distinguished farmer, soldier, statesman, and patriot, Washington; who had collected at Mount-Vernon whatever he believed useful to the agriculture of his country; and, among other animals, a Persian ram, which Mr. Custis describes as being very large and well formed, carrying wool of great length, but of a coarse staple. This stock, intermixed with the Bakewell, are the source from which the fine Arlington sheep are derived; some of which, he says, carry wool fourteen inches in length, and are formed upon the Bakewell model. I have never seen these sheep, but from Mr. Custis's description, and from the produce of the wool at the public shearings, I have no doubt that they are a valuable race, and such as merit the attention of those whose farms yield a good rich bite of grass; for upon any other I would never recommend long-woolled sheep. The sample of wool which Mr. Custis sent me from this stock possessed every ingredient which is esteemed in combing wool. It was fine for the sort, soft, silky, and beautifully white. It is admirably calculated for hose, camblets, serges, and other fine worsted fabricks; and it would be a pity to see it diverted to any other objects, or to the making of fine cloths, for which it appears to me less adapted. It is, however, matter of surprise, that a Persian ram should be the parent stock from which this valuable breed is derived. The wool of Persia has always been considered as among the finest in the world; the white sells nearly upon a par with that of

Spain at the London market, and the red somewhat higher. Either there must have been some mistake as to the place from which the ram came, or Persia must possess two distinct breeds of sheep: indeed, it is not improbable that the southern parts of Persia, upon the Indian Ocean, and Gulf of Ormus, may contain the large coarse-woolled sheep that are commonly found in Africa. For the Smith's Island wool we are also indebted to the researches of Mr. Custis; from whose valuable pamphlet I have extracted the following account of it:

" I come now to speak of Smith's Island wool, " a discovery from which will arise the happiest " effects to my country, and yield the most " grateful sensations to myself. This island " (which is the property of Mr. Custis) lies in " the Atlantic Ocean, immediately at the east- " tern cape of Virginia, and contains between " three and four thousand acres. The soil, " though sandy, is in many parts extremely " rich, and productive of a succulent herbage, " which supports the stock at all seasons. About " one half of the island is in wood, which is " pierced with glades running parallel with the " sea, and of several miles in extent. These " glades are generally wet, and being complete- " ly sheltered by the wood on either side, pre-

" serve their vegetation in a great measure " through the winter, and thereby yield a sup- " port to the stock. Along the sea coast are " also abundant scopes of pasturage, producing " a short grass in summer, which is peculiarly " grateful to the palate of most animals, and " particularly so to sheep. The length of this " island is estimated at fourteen miles, which " gives that variety and change of pasture so " necessary to the system of sheep farming. " Within it are various shrubs and plants, which " the animal appears to browse on with great " relish, particularly the myrtle bushes, with " which the island abounds. The access to " salt also forms a material feature in the many " attributes which Smith's Island possesses.

" The origin of the Smith's Island sheep can- " not be precisely ascertained, but they are " supposed to be the indigenal race of the coun- " try, put thereon about twenty years since, and " improved by the hand of nature. When we " compare Smith's Island wool with the native " wool of the country at large, we are lost in " astonishment at the wonderful interposition " of providence in our behalf, which serves to " show what benefits we enjoy, and how little " we have estimated the gifts. The Smith's " Island wool is, without question, one of the

" finest in the world, and has excited the praise " and astonishment of all who have seen it. To " recapitulate the various opinions given of its " merits is unnecessary. It only remains to be " judged in Europe, whither a specimen has " been sent, to determine its value when com- " pared with the famous Merino, hitherto the " unrivalled material in the woollen manufac- " ture. The Smith's Island is a great deal " longer than the Spanish, being in full growth " from five to nine inches in length, and in " some instances more. In quantity it is also " vastly superior, as the sheep yield twice as " much, and in some instances more. And, " lastly, the size and figure of the animal ad- " mits of no comparison, being highly in favour " of the Smith's Island. The only remaining " question is the texture. If the Merino is " finer in grain, the Smith's Island is so fine as " to answer every purpose to which the other " can be appropriated, and so much larger in " quantity as to yield a better profit to the bree- " ders. No cloth which the Merino manufac- " tures will be disgraced by the introduction of " the Smith's Island; and many fabricks ma- " nufactured by the one at a great price, can be " manufactured of the Smith's Island at much " less. The Smith's Island is as white as snow, "and perfectly silky and soft to the touch, and "of delicate grain." Mr. Custis adds, that these sheep are shorn twice a year. I have written to him to know why that uncommon mode of shearing has been pursued. He informs me that he would inquire and answer my query; but I have not yet been favoured with the information I wish on this important subject.

Mr. Custis not having mentioned in his pamphlet the quantity of wool shorn at each time, I am enabled in part to do it from one of the letters which he has done me the favour to write me. He says the *best* of these sheep have yielded four pounds at a shearing, making an aggregate of eight pounds per year. It appears to me Mr. Custis is not fully informed either of the fineness or the quantity of wool produced by the improved Merino, when he supposes that eight pounds of unwashed wool from the *best* of his sheep is more than double the produce of the Merino. I have shorn from one of my Merino rams of the improved French breed, eight and an half pounds of unwashed wool, and from another seven pounds and three quarters; though my pastures being extensive, my sheep, kept free from filth in the winter, are remarkably clean when they come to be shorn. In France, from twelve to thirteen pounds is said to be the average fleece of the rams form the national flock; but then their sheep are very dirty, for the reasons I have mentioned. I should also add, that the price of my Merino wool has risen from one dollar and twenty-five cents, at which I sold when I wrote to Mr. Custis, to two dollars. Since the hatters and clothiers have examined its texture, forty-four and one-half yards of fine close-wove cloth, forty-five inches wide, as it came from the loom, have been made from sixteen pounds and three quarters of it.

I have given these extracts from Mr. Custis's valuable pamphlet, because I think it important that the country should know its resources, and be enabled to select a stock adapted to their soil and to their wants. I cannot, however, agree with him in sentiment (as far as I can form mine from the sample he sent me), that the Smith's Island wool can be introduced into any of the manufactures in which the Merino wool is used. It is soft, white, and silky, but neither so fine or soft as the Merino wool. Mr. Custis has, however, taken the proper method to ascertain its value, by sending samples to Europe, and will, I trust, furnish the public with the result of his inquiries. I cannot omit this occasion to express the high opinion which I, in common with every other person, entertain of Mr. Custis's patriotism, and of his animated exertions for the improvement of this most important branch of our rural economy.

There still remains a breed of sheep to be noticed, which might indeed more properly have been mentioned before—the Thibet or Cassimere sheep. These are said to carry finer wool than those of Spain; but from their remote inland situation they are little known, though I think I have been informed that one was brought into England either by Lord Cornwallis or the Marquis of Wellesley. We may form some judgment of the fineness of their wool by the shawls that are imported from India, and which we have whimsically called camel's hair shawls. These fine cloths are made for turbans, and are of two sorts; the finest, I believe, never go out of India, as we may judge by comparing those we meet with to Taverner's account of one presented to the Grand Mogul of sixty yards in length, which was folded in a cocoanut shell. The best are made from the wool plucked from the breast of a wild animal which is not particularly described, but which probably, as it is a native of the mountains, is either the Vigone or some animal of the same species. The other, which compose the finest exported from India, is made from the wool of the Cassimere and little Thibet sheep, these countries being in the vicinity of each other.

Having taken a cursory view of the different breeds of sheep, which I conceived would afford matter of amusement to my readers, and perhaps lead to deductions useful in the improvement of the breed, which, however, I shall not attempt to make at present, I will proceed to such practical observations as may be found useful to those who have given less attention to the subject than I have done, or who have not the means of knowing what more experienced farmers have written on the subject.

CHAPTER II.

THE United States of America, particularly those which lay to the north of the Chesapeake, appear to me to possess advantages in the breeding of sheep which are unequalled by those of any part of Europe which I have seen. First, the country is generally hilly; the hills covered with a fine herbage; almost every pasture is furnished with running water, and sheltered more or less by trees against the summer sun; the enclosures are much more extensive than those which are found in the few enclosed counties of Europe; where, except in England and Holland, scarce an enclosure is to be seen; and in these countries they are so small as to be ill adapted to sheep; which, on that account, very generally run on commons. Where there are no enclosures, the sheep must necessarily be folded at night at all seasons of the year, a practice extremely hurtful to them. Again, when they are turned out, they must be led over fallow grounds, or pick the scanty herbage upon exhausted fields. They must be surrounded by shepherds and their dogs, to prevent their trespassing upon the crops, which have no other protection: they must, of course, be kept in such close order as never to be without the atmosphere of each other's breath. What wonder then is it that our sheep are subject to few or none of the diseases that so frequently diminish the flocks of Europe? It is true that Spain may be considered as forming an exception to what I have said; not because of any natural advantage that she enjoys, other than in having made a happy selection of her flocks; but because the whole agriculture of the country has in some sort been sacrificed to the maintenance of their sheep, as I have already stated in the preceding chapter. The price of wool also in this country is, in proportion to the quality, higher than in any part of Europe, while the value of land is much lower. The interest then of the farmer with us unites with his patriotism in calling his attention to the improvement of his flock. In doing this, the first object must be to adapt his breed to his soil and situation. If he lives in the vicinity of a great city, whose wealthy inhabitants will be less mindful of the value and the price than of the rarity of an object, let him adapt his flock to the demand that their taste, their whims, or their luxury may make upon it. His early lambs will in this case bring such a price as to make it an object to keep a breeding flock of that species of sheep which will produce the earliest lambs. The most celebrated stock which I know for this purpose is the Dorsetshire sheep, from which are bred the house lambs which supply the London market. The ewes are kept in high order, and are put to the ram in the months of May and June. The lambs are fit for market during all the winter months, and on that account bring an extravagant price. They are kept in the house at all times, and the ewes turned out, but brought in to them at night and at noon. A lump of chalk is given to the lambs to lick, which is said to make the meat white. When a ewe loses her lamb, or it is killed off, she is compelled to admit another, and is held if she refuses it. The lambs by this means having both a mother and a foster-mother, are rendered sooner fit for the butcher. The ewe is kept upon the most succulent food while she gives milk; but it is a rule among the breeders to keep the earliest ewe lambs for stock, and it is probably an attention to this circumstance that has produced a kind of sheep that will take the ram at so early a period. By the same attention perhaps, and by keeping both the rams and the ewes very high, a similar breed might be made among us, if the original Dorsetshire stock could not be procured; though in time of peace there is very little difficulty in obtaining them from England by means of the smugglers that trade from Dunkirk, notwithstanding the high penalty which the government has very ungenerously annexed to their exportation.

If the farm on which sheep are to be reared consists of wet or marshy ground, with rich and luxuriant grass, I would recommend that the large sheep, bearing combing wool, should be preferred, since the largeness of the carcase and the quantity of the wool might, in such ground, more than compensate for a diminution in the price of the wool, if the scarcity of such wool in our country should not (as might be expected) enhance its value: in fact, we have hitherto made very little distinction, and we sell alike wool that in England would bring but twelve cents, and that which in that manufacturing country would be valued at thirty-six cents. Hence, where the pastures are adapted to large, long-woolled sheep, they would for the present be highly valuable, and particularly in the vicinity of the sea; for I have observed, that the

English sheep which have been introduced into this country degenerate much less on the sea coast, than when they are conveyed beyond the first ridge of mountains. England, Ireland, and Flanders will supply the stock, if it should be thought that those offered to the public by Mr. Custis should not fall within my description. I should, however, both from his account of the Arlington long-woolled breed, and from the sample I have seen of their wool, think it unnecessary to look further for a stock adapted to the pastures in question. Their size will increase with their pastures, and the length of their wool with their size. For every other description of pasture I think no doubt can be entertained of the preference that should be given to the Merino breed. These may be found of such size and constitution as are adapted to any ground. Those that are dry and barren, such as our shrub-oak plains, will find in the small Merino, which are common in most parts of Spain, a stock which will not only subsist, but thrive on such grounds; and though their fleeces are lighter, they are not less fine than those of the larger and more improved breeds. The faults in their form will, by an attentive breeder, not fail to be gradually corrected. From Spain may also be procured, by those who have the means of selecting and will not spare expense, a larger breed, with heavier fleeces and better forms, and with equally fine wool. I have now in France a few that have been so chosen in Spain, for which a double price was paid, and which are of uncommon size and beauty: with these, and a number more from the first flocks of France, I hope to enrich our country when means shall be afforded for bringing them out.

From France the best stock may more easily be obtained,* and being already acclimated to that country, which is more similar to our own, and used to be fed on hay, and not to migrate, there will be less risk in the importation and in the adapting of them to our climate and manner of keeping. Of these I have already treated. I proceed to state my ideas on the best and cheapest mode of obtaining a Merino flock, or such a portion of Merino blood as shall instantly double the value of a flock of sheep. The high price of Merino rams, and the difficulty of procuring those of the best sort, will deter many farmers from entering at

* Colonel Humphreys, who has probably seen both those of France and Spain, agrees with me in this sentiment. In his letter to the Agricultural Society of Massachusetts, he states, that the improved stock of France yield twice as much wool as those of Spain, without any change in the quality of the fleece.

once upon the enterprise in the most effectual manner; that is, by procuring full-blooded rams in the first instance: and as this essay is not intended for those whose wealth enables them instantly to overcome all difficulties, I shall treat the subject upon so economical a scale as to be within the means of every man that keeps a flock.

After having determined on the kind of sheep most proper for your farm, which we will suppose to be Merino, carefully examine your ewes, and select from them those that have the shortest or thickest coat, with the least hair on the hinder parts, and whose bellies are well covered with wool. Those whose wool is neither long enough to comb, nor yet so short as to be good carding wool, should be immediately sold or exchanged for others of the description I have mentioned. In this there will be no difficulty, because, generally speaking, they will be the largest; and as their long wool covers their defects, they apparently are the handsomest in the flock. Let your ewes be at least three years old, as large as can be got of the sort, belly large and well covered with wool, chine and loin broad, breast deep, buttocks full, the eyes lively, the bag large, and the teats long. Next provide yourself with a ram possessing as much of the Merino blood as you can conveniently afford to purchase; let us suppose him to be half-blooded. In choosing him, be particularly attentive to his form and size, that you may not diminish, but rather add to the beauty of your flock. Let him be broad in the chine and loins, deep in the carcase, the back straight and neither arched or swayed; the ribs set out so as to afford room for a large belly well covered with wool, the forehead broad, the eyes lively (for a heavy eye is the mark of a diseased sheep), the testicles large, and if covered with wool it will be an evidence of his taking after his sire; let him be strong, close knit, and active. To judge of his vigour, take him by the hind legs, and observe if he struggles with force, or makes but a feeble resistance. Next, as the most essential point, examine his wool; if it is as fine as you can expect in a sheep of his grade; if it is thick, close, and greasy, full of yoke, and the breast and loins also well covered with wool, you may rely upon his goodness. Upon the thighs of a sheep of this grade you must expect to find more or less coarse wool; if, however, you have the means of choosing, take one that has least of it. I should prefer making my stock gradually in this way, out of well-formed, good-sized rams, to attaining more blood at the expense of size and beauty; for though size may be in itself of minor importance; yet if you afterwards attempt to increase it by larger rams, you will find some difficulty in doing it when your stock of ewes are small. They will lamb with more difficulty, and afford less milk in proportion to the size of their lambs. Beauty of form is always to be considered; for the best formed sheep are generally the most thrifty. Such a ram, with the ewes I have described, will give you one-fourth breed lambs, who will carry at least one-fourth more wool than your old stock; and this wool will not be worth less than fifty-six cents the pound, if that of the ewes sold at thirty-seven cents. The quality and quantity of the wool taken together will nearly double the value of your fleeces in the first generation.

Now let us see at what expense this advantage is purchased. The ram we will say cost twelve dollars. The first year he will give you, if well kept, and not exhausted by too many ewes, five pounds of wool, worth one dollar per pound as wool now sells; charge his keeping at one dollar and fifty cents—clear profit, three dollars and fifty cents, that is, 33 per cent. on his original cost; so that instead of paying any thing for a ram which shall double the value of your flock, you have only put twelve dollars to a more advantageous interest than any other stock would have afforded. Any farmer then who can raise the money, either by borrowing or parting with some of his other stock at even something less than its value, to procure such a ram, must stand greatly in his own light if he hesitate about making the purchase, because the returns are great and certain. Suppose his original stock yielded him thirty-six pounds, from which must be deducted the keeping, which will absorb the whole, his new stock being one-fourth breed, will, in the increase and fineness of the wool, add at least thirty dollars more to it. Thus, for twelve dollars expended, he receives in eighteen months, when his lambs come to be shorn, thirty-three additional dollars, and two fleeces from his ram, worth nine dollars more, and this all clear profit beyond the keeping of his sheep, which the old fleeces would but just have paid. Is there any farmer so blind to his interest as to breed any longer the common sheep of the country, when his flock may so easily and so reasonably be renovated? But he should not stop here: the clear profit upon his flock after the first year, and the price of his ram, which he should then sell, will enable him to purchase a three-fourth blood ram, say at twenty-five dollars. Such a ram, with his one-fourth breed ewes, will at once give him a half blood flock, and that without any expense, because he purchases him with the excess beyond what his original stock would have given him.

It is a general practice, and I believe in most cases a good one, to let the lambs come in April; but in changing a stock I should prefer a different course, though it may be attended with some more trouble and expense in taking care of the ewes. The lambs that come in April will take the ram, but their young will be feeble, and even if they should live, will not form a good basis to work the change upon; I would, therefore, put the old ewes intended for stock in good heart, and kill off the lambs early, so that they may take the ram in August and September. The lambs that fall in January and February will be large and strong enough to put to ram the November following, and produce good lambs, so that a year may be gained by changing the flock. It is true that it would tend to the improvement of the flock not to let the ewes breed till they are two years old; but few farmers have the means to keep them away from the ram, and they will generally take him in November and December; in which case it is better that the ewes should have acquired two or three months more growth. With a very large stock this might be troublesome and hazardous; but a farmer that keeps about thirty ewes might do it with little loss or inconvenience, by raising a few cabbages, turnips, and potatoes extra, to gives his ewes at yeaning time. I presume that every farmer knows that a ewe goes five months with lamb, and of course how to regulate the yeaning by the keeping off or admission of the ram. The number of ewes that a ram will cover has never, that I know of, been precisely ascertained. The Spanish shepherds have one to twenty-four ewes, and this seems to have been the rule in the days of the patriarchs, as we may infer from some passages in which their flocks are enumerated. In France they seem to think forty the most common. In England a ram highly kept has gone to eighty ewes, but then precautions were used to keep him from exhausting himself, by giving him only one at a time. Without these precautions, however, I have generally found one ram sufficient for sixty or seventy ewes; and have even known one to serve a hundred, but I think he was injured by it. If I had my election, I should not choose to put more than forty ewes to a ram. If the rams are let to run the whole season with the flock, one will serve more ewes than if kept apart till late in the autumn.

I should suppose that every good farmer would provide some shelter for his ewes in the winter; if he does not, he ought by no means to let his lambs drop early, or he will meet with great losses.—It will be proper here to mention the manner in which I think the flock should be treated in the winter, which the attentive farmer will either adopt or improve upon, as circumstances may demand. It is common to let the sheep run about the barn door, and pick up what the cattle drop. This may be economical, but is not a practice calculated to make a fine flock. The sheep will frequently be hurt by the cattle, and the timid ewes will be driven from their food. It might not be amiss to have a few wethers kept for that purpose, who would probably winter well with little expense by running with the cattle. The stock wethers, if the flock is large, should be kept by themselves. If they run in an open field, in which there are hills, trees, fences, or houses, and are foddered from the hay-stack, they need no other shelter; though I should prefer a rick out of which they were fed, and on the windward side of which they could lay. If they run with the ewes, as they are stronger, they will feed upon the most delicate hay, and compel the ewes to eat the refuse. They will also render it difficult to give the ewes separately that succulent food which they require before and after they have lambed, unless you are provided with such a quantity as will serve for the whole flock, both ewes and wethers. The following is the plan I pursue for a stock of two hundred ewes and seventy wethers.

I have chosen two warm dry situations, sheltered from the north-west wind by hills, and open to the morning sun in winter at its first rising. On the north side of this I have erected two barracks of about twenty-four feet square, with an elevation of about six and an half feet from the ground to the hay-loft. These, standing at a distance from each other, I have united by a shed having the same elevation, and being about ten feet deep, with a hay-loft above. This shed is open to the south, and boarded to the north; the barracks are boarded up, the one on the north and west, and the other on the north and east; the sheds cover the east side of one and the west side of the other, uniting them together. Along the whole of this building racks are erected, with a trough at the bottom to catch the hay-seed, of which the sheep are very fond. This trough also serves for turnips, bran, salt, &c. and as the extent is accommodated to the number of sheep, they are equally fed, the strong having no advantage over the weak. On the outside of this building all around, are boards hung upon hinges, which serve to put the hay in which is thrown from the barracks to the outside of the sheep-fold. By this means the wool is kept free from hay-seeds, which injure it very much. Along the racks, for the distance of seven feet, the building is floored, so that the sheep are kept clean and lie dry. The yard is about three-fourths of an acre, and is surrounded by a high pale fence, that dogs may find no admittance. In warm days, when the sheep are out, the loop boards along the rick are turned up, so as to let the wind pass freely under the studs, and render the air fresh and pure. With these buildings I am very little solicitous about keeping away the rams till late in the autumn. My lambs generally come in March, and sometimes earlier, and by having an attentive shepherd, I have seldom lost many, even when, as the year before last, by the severity of the weather late in March, my neighbours lost many of theirs.

The following is the practice I would recommend, founded on my own experience, if the lambs come early; and I cannot help thinking that those that do, winter better the ensuing year, and make the finest sheep, at least if the ewes are suffered to breed the first season. In France, however, they are ever attentive to keep the Merino ewes from taking the ram till two and an half years old; and to this circumstance among others, they attribute the great improvement of the stock. Indeed, a full breed Merino will not take the ram till she is eighteen months old, at least this has always been the case with mine. After having provided shelter to which your ewes may retire in bad weather, care must be taken to furnish the yard with a great quantity of litter, and to renew this after every rain. This furnishes a quantity of manure that richly repays the expense of the litter; it keeps the wool clean, and contributes greatly to the health of the flock; if your lambs are to come early, it is still more necessary, since without it many will be lost by dropping during a wet or cold night upon the damp ground, to which they sometimes freeze; and the filth which they by this means contract, will often keep the ewe from licking them dry. I generally heap up leaves (which I collect in the autumn) about a foot deep, and occasionally lay straw upon them. This forms a soft bed in the winter, and by its early fermentation in the spring, furnishes a rich manure. In stormy weather your shepherd should visit your fold very frequently about yeaning time, as a storm appears to accelerate the birth of the lambs, and some may be lost for want of attention.

In addition to the general fold, I have four partitions under the shed, large enough each to contain a couple of ewes. When a lamb drops, it is put, with its mother, into one of these enclosures, which is well littered. Here they are kept for two days, and the ewe is fed with bran and succulent food. When more lambs come, and these cells are wanted, the older give place to the younger, the lamb being generally sufficiently strong the third day to take care of itself, and to find its dam when turned into the flock. In the early part of the season, and before the ewes begin to show any signs of being near their yeaning time (which may be known by the swelling of their udders), they are kept upon good hay (clover is preferred to any other) and corn-stalks. When any of them appear to make bag, as the shepherds call it, which will be about ten days or a fortnight before they lamb, they are carried to the second of the sheep-folds that I have mentioned, and are there fed with the best of hay, corn-stalks, turneps, cabbage, or potatoes, and once or twice in a day have a handful of wet bran. This gives them a flush of milk when the lambs drop; for want of which many lambs are lost by inattentive farmers. In this fold the lambs and ewes are kept separate from the rest of the flock, till they amount to about half the number; when those in the first fold will be so far advanced as to require the same treatment, and are so diminished in number as to make any removal unnecessary, the whole stock being then well fed with the most succulent food that can be procured for them. Whenever the snow is off the ground they should be turned to pasture, with the exception of those whose lambs are too young to follow them; and even when the snow lays, if not too deep, they should be led out to water; and if you have any cedar, pine, hemlock, or other bushes that rise above the snow, it will be well to beat a path to them, and leave your flocks an hour or two among them. The branches of these trees too should be brought into the fold when the ground is long covered with snow; for the feeding on them will contribute much to the health of your flock. Where these cannot be obtained, smear tar on boards, and sprinkle them lightly with salt, and lay them so as the sheep may get at them; by eating it their bodies will be kept open, and themselves in heart. Once a week a small quantity of salt should be given in the mangers. Salt is, I think, essential to the health of sheep in our climate, and it is thought of so much consequence in Spain, that the King cannot raise much revenue on that article, lest it should induce the shepherds to abridge the quantity usually given to their sheep; which, they say, would not only injure them, but change the quality of the wool. About a fortnight after the lambs drop, give them, besides your mark of appropriation, a mark which is to distinguish the degree of Merino blood they possess. At this time you can make no mistake, because you can take the lamb immediately from its dam: if you defer it till they are larger and more numerous, you will be liable to errors. I view this as very important, particularly if you mean to sell any of your stock for breeding, since a man that possesses either honour or honesty would feel the utmost pain at having deceived the purchaser in a matter which is so essential to the amelioration of his flock.

Should any deformed or lame lambs be found

in your flock, or should any one be killed by accident, strip off the skin from such lamb, and cover with it either a twin lamb or the lamb of a young ewe who does not appear to be a good nurse, and shutting up the ewe that has lost her lamb, she will generally take it as her own. Should she refuse, she must be held for a day or two, when she will adopt it. This is a common practice in Spain, where even half the lambs are killed, and two ewes given to each lamb. The fatigue they undergo in travelling I presume has rendered this necessary. One of my neighbours tried it last spring, upon my recommendation, with success. If the lambs come early, the ewes will be relieved, and the lambs strengthened by giving them fine hay and bran, or any succulent food, such as cabbage, &c. In order to do this, and not suffer their food to be eaten by the old sheep, I have contrived boxes with a rack and manger within them, and lids to put in their fodder. The front of this box is of lath, so wide as to permit the lambs to go in and out at pleasure, but too narrow to admit the grown sheep. If it is preferred to have the lambs come in April, in that case no particular care is necessary, other than that of providing a field of rye or clover for the ewes.

Having brought our flocks through the winter, we now come to the most critical season, that is, the latter end of March and the month of April. At this time, the ground being bare, the sheep will refuse to eat their hay, while the scanty picking of grass, and its purgative quality will disable them from taking the nourishment that is necessary to keep them up. If they fall away, their wool will be injured, the growth of their lambs will be stopped, and even many of the old sheep will be carried off by a dysentery. To provide food for this season is very difficult; turneps and cabbage will rot, and bran they will not eat after having been fed upon it all winter; potatoes, however, and the Swedish turnep called the roota baga, may be usefully applied at this time, and so I think might parsneps and carrots. But as few of us are in the habit of cultivating these plants to the extent which is necessary for the support of a large flock, we must seek resources more within our reach. The first and simplest of these is to leave the second growth of our clover uncut, and to turn the ewes upon it. The young clover will shoot very early in the spring, having been covered by the old crop during the winter. This, together with the old grass, which the sheep will be compelled to eat in order to get at the young sprouts, will keep them up till the pastures are fit for them. A still better practice is to put in a very early crop of rye, giving the ground a double quantity of seed; and perhaps too if the seeds of turneps, kale, and winter cabbage were sown with it, they might, if the winter was favourable, add to the quantity of food. The ewes and lambs turned upon this would thrive exceedingly, and if your pastures consisted of rye grass, orchard grass, clover, parsley, and burnet, which come forward early in the season, they might be taken from the rye before they had done it the least injury; their feet and tails more than compensating the mischief done by their teeth. The summer feeding of sheep must of course be regulated by the nature of the owner's ground; if, however, it is in his power to make a selection, let him choose grounds with a sweet herbage of white clover, spear-grass, or blue-grass; let the pasture possess both water and shade; and as sheep prefer short grass, and have no objection to feeding after horses, though they dislike what other sheep have lain or breathed upon, it will be economy to put horses on the same pasture; horned cattle are not good, because ruminating animals dislike the food that is tainted with the breath or tread of other animals that ruminate. It will be proper too occasionally to change the pastures. I find that the daisy is eaten readily by sheep in the pasture in the spring of the year, and when in flower they will crop the flowers; but then they must have a change of food, or they will tire of it, and only eat it from necessity. If not used to it, it will sometimes purge them in the spring; in which case their pastures should be changed. No hay is, however, eaten with more avidity, both by sheep and cattle than that made from the daisy when in the flower. If it stands thick, and you cut it down, after wilting a few hours the cows will leave their grass to feed upon it.

It is a generally received opinion in every part of Europe except England, that sheep should not feed either in the evening or in the morning when the dew is on the grass. Nothing can be more absurd than this idea, or more contrary to experience. With me it is one among a thousand other proofs, that fraud may practise upon ignorance till falsehoods are considered as the axioms of truth. In every country in Europe except England and Holland, sheep are tended by shepherds, who lead them to the field, and continue out with them the whole day, whatever may be the state of the weather. It was very natural for men who had no interest in the prosperity of their flocks to endeavour to abridge this wearisome and lonely task, to share early in the evening the pleasures of society, and enjoy their fire-sides, and to quit their homes as late as possible in the morning. This well-invented tale answered their purpose, and, perhaps, in the beginning derived force from the accidental or fraudulent death of some part of their flocks. The shepherds were too much interested in supporting this idea, and their masters too ignorant, or too confident in their integrity to refute it; and from hence this system of keeping up the flock till the dew dries off the ground is so general as never once to be doubted in every country where flocks are tended by shepherds, and ridiculed in those in which they feed without a guard. In England sheep are out night and day. In America the sheep are found to feed with most avidity when the dew is upon the grass. If the pasture is plentiful, they fill themselves and lay down by nine o'clock, and rise again to feed an hour after; but as soon as the sun has perfectly dried the grass, and began to beat upon their heads with violence, they seek the shelter of some friendly shade, and will even suffer hunger rather than take their food while they may be incommoded by the heat. If the pasture affords a wood or a hill, under the shade of which they can feed, they will be found on their legs again by three or four o'clock in the afternoon; but if not, they begin to feed later in the day, and will continue so to do some hours after sunset. It will easily be conceived then, that sheep must suffer extremely by being folded when they should feed, and being compelled to feed when they should be at rest.

I ought to have mentioned, that it is a practice in some places to shear the tags and wool from the udders of the ewes before they lambed; and this practice is strongly recommended by a number of agricultural writers, who allege that the lamb cannot suck so well unless this is done: but there are many plausible theories which are not confirmed by practice, and this I take to be among the number. The teat is always bare, and this is the only part that the lamb has any thing to do with, and bareing other parts only tend to mislead his search. But this is not the greatest evil that results from it. The ewe must be handled, and too often very roughly, when she is heavy with lamb. The effect of this is very obvious; the teat is sometimes wounded by the shears; but, above all, the shearing exposes the udder to cold, which, if the ewe is very forward, throws back her milk, and sometimes kills her; and even when less forward, it endangers her health, and of course that of her embryo lamb. I have seen an account, though I cannot just now recollect where, of a number of ewes dying in England in consequence of cold weather following soon after this unnecessary operation.

SHEARING.

This too is a delicate task, and requires more attention than is frequently paid to it. Many farmers begin by washing their sheep. This may be a good practice where the fleeces of the sheep are thin and shaggy, as in the long-woolled breed; yet I think with thick, short clothing wool it is of little use, and particularly if the flock consists of Spanish sheep, whose wool is so close and thick as to render it absolutely impossible to make it clean by washing on the sheep's back; and for this reason it is never practised in Spain. The long, straight wool soon dries, and therefore the sheep are less injured by it. But when the water is made to penetrate to the skin, through a thick close fleece, it will remain wet a long time, and I think cannot fail to injure the sheep, which are very subject to colds in the head, chills that penetrate the limbs, and, falling on the bowels, bring on a lax, which sometimes kills, and never fails to weaken them extremely. Another evil which is little attended to, is the bringing together a large flock of sheep in a stable or close barn, and keeping them together till the whole are shorn. If there are any disordered sheep in the flock, they communicate their complaint, if contagious, to the whole flock, who take in each other's effluvia at every breath they draw. But independent of this, their being heated in this manner, and immediately after stripped of their clothing, cannot but be very hurtful to them. In Spain it is a common practice to keep the sheep closely confined, in order to make them sweat, with a view to increase the weight of the wool, and to make the shears enter easier. The consequence is, that many die; and in some instances one half of the flock has been carried off in the space of a night. I cannot but believe that this injudicious management and folding have generated that great catalogue of maladies that prevail among the sheep of Europe, but most of which are happily unknown in America. I would, therefore, recommend, when the shearing commences, that the sheep be penned in the open air, and brought by six or eight at a time into the barn. If the flock is large, drive up only one portion of them, and let the rest feed abroad till wanted. The time of shearing must be regulated by the state of the weather and the growth of the wool. If the sheep begin to loose their wool, and this does not arise from bad keeping, it will be found, on examination, that it is protruded by a growth of young wool; there would then be some loss by deferring the shearing, as the new wool will injure the old, and the next year's crop be diminished in quantity by the delay. But even this should not induce the farmer to shear his sheep till the weather is warm and settled. In this circumstance the Merino breed have an advantage over all others. They never shed their wool; and from some experiments that have been made in France, it appears that two and even three years growth may be had at one cutting without diminishing the quantity. Thus, if a sheep would have given three pounds the first year, if left unshorn, he will give six the next, and nine the following; so that if it was desirable to have Merino wool of ten or twelve inches long, it could be obtained: but it is a practice that I would not recommend in our warm climate, where sheep must suffer greatly under so thick a fleece, as well from the heat as from the lice that it would generate. It is, however, a great advantage not to be compelled, from the falling of the wool, to shear at an inconvenient or improper time; and this advantage is, I believe, confined solely to the Merino breed. How far it may extend to the mixed breed I do not know.

In some countries the sheep are shorn twice a year; but wherever this practice prevails, I believe it is owing either to the wool's being too coarse for use when it attains its full growth, or because, as the winter approaches, and no proper provision is made to keep them, the sheep falling in flesh, would not keep their wool till shearing time.

It is a general practice in shearing to tie the legs of the sheep together. This is very improper: it forces the sheep into a position in which the intestines being pressed, they discharge their urine and dung at the time they are sheared, which fouls the wool, and is offensive to the operator; besides which, the skin being by this means drawn together, there is much more danger of cutting the sheep than if they were placed in their natural position. It has, therefore, been recommended to tie them to a table, after laying them on one side; but this I think would subject them to some risk if they should struggle, and at all events will require twice tying, as the sheep must be turned. I contemplate trying the next year the tying the fore and hind legs to a bar wtih two cross pieces; the bar to be about eighteen inches long, and the cross pieces six. This would leave the sheep in their natural posture, with their legs a little stretched out; a rod of iron, with a curvature at each end, would perhaps be still better, because, being smaller, it would be less in the way of the shears. The shearing of a common long-woolled sheep is a matter of little difficulty, the fleece being light, and the wool not so valuable as to occasion any great attention to shearing close. To those that have short-woolled, and particularly Merino sheep, I would recommend not to trust the shears to careless hands, or by any means to hurry their workmen; on the contrary, to remind them constantly that the wool is sufficiently valuable to compensate for the time spent in taking it off, and the sheep too valuable to be maimed. To shear one Merino sheep properly will take more time than to shear three long-woolled sheep. Let the master then show no impatience if he would have his work well done. Great care should be taken not to wound the sheep, particularly when the shears is applied near the udder, where wounds are dangerous; but as some accidents will unavoidably happen, the best remedy to apply to the wound, in order to heal and protect it from the flies, is a little tar from the tar-bucket, which contains some mixture of grease, and a little fine dust of charcoal over it.

If the wool is to be used in the family, it is best to sort it as the fleeces are taken off, putting the wool of the hoggets (or young sheep) by itself, because it injures cloth to mix this with that of full-grown sheep, as it is not of the same texture or strength, and will make the cloth shrink unequally. The other fleeces may be sorted also, making separate parcels of the thighs—the belly—the back and sides. Another assortment may be made afterwards if thought necessary. If the wool is designed for sale, the fleeces should be carefully rolled up, first taking off the tags, and tied together with a lock of the wool. If the flock consists of Merinoes, pure or mixed, and common sheep, as many parcels should be made as there are grades in the flock, so that each part, when carried to market, may be marked according to the value. The wool should not be kept long upon hand without washing, as it is liable in that case to ferment and spoil in hot weather. When the sheep are shorn, if the weather should prove wet and cold, and you have sheds or barns of sufficient size to contain them without crowding, it will be well to house them at night, and to give them salt, which is a stimulant, and will enable them better to bear the sudden chills occasioned by the loss of their fleeces. To this might be advantageously added a little corn or oats.

Before the sheep are dismissed from the shepherd's hands, they should be carefully examined as to their age, their constitution, and the quality of their wool; the old sheep, those that are weak, ill formed, or ewes that appear to have been bad nurses, or to have lost their lambs from the want of milk, or whose wool is bad, either by being mixed with jar (short hairs), or which are rough on the thighs, should be marked, in order to turn them off, and put in good pastures by themselves, to fat them the sooner. The age of a sheep is distinguished by their front teeth; they have eight in their under, but none in their upper jaw. These are complete at their birth, but they are small and pointed. The second year the two middle teeth are changed for two of considerably more breadth, by which they are distinguished from the six lamb's teeth: the third year the two adjoining teeth are changed: the fourth year leaves them with only two lamb's teeth: the fifth year all their teeth are changed, and they are then said to be full mouthed. At seven and eight years they begin to loose their front feeth. Whenever this happens, they are called broken mouthed, and should be turned off to fat, as they are then upon the decline. Many sheep will, however, preserve their teeth much longer; but it will be best, except in the case of a valuable ewe or ram, to turn them off the year after they are full mouthed; for they decline in wool as they grow older, and will fatten better before the teeth get loose or decay: besides that, in a flock of common sheep the principal profit consists in fatting off early, as the sheep sold is of three times the value of the lamb by which it may be replaced.

The shearers should also examine very attentively whether the sheep they shear have the scab, which may be known by the wool coming off easily, and by the skin being rough and discoloured. In that case the remedies which I shall hereafter mention should be immediately applied; and if the disorder has gone so far as to have formed a sore or scab, it will be best to separate the infected sheep from the flock till the cure is effected. Cold and heat is injurious to sheep that have been just shorn; they should, therefore, be put into pastures in which they can find shade, for the sun not only hurts them when naked, but dries the skin, injures the growing wool, and is said to produce the scab. At this time too attention should be paid to the horns of the sheep, to see that they do not press upon the skull, or endanger the eyes; in either of which cases, if not taken off, they will cause the death of the animal. There are two ways to remove the horns: the Spanish shepherds make a hole in the earth large enough to contain half the body of the sheep, and another of less depth for his head, under which a block is placed; the animal is laid upon his back in the pit, a man sitting astride to keep him down, while another confines his head to the block, and a third cuts off his horns with a sharp chissel and mallet; I have, however, preferred using a fine stiff-backed saw, with which the operation can be very neatly performed, though a surgeon's saw would be better. After the horns are taken off, I have applied tar to the extremities of the stumps, and tied over them two or three folds of strong linen to keep off the flies. Last year I was compelled to have this operation performed in the heat of summer, and to take off the horn, which was very large, within two inches of the skull; and though it bled freely, the ram did very well, and seemed not to feel any inconvenience from the operation after the bandage was applied. It would be desirable to obtain sheep without horns, not only to avoid the trouble of cutting them, but because they are dangerous to the ewes in the fold, inconvenient to the sheep if they feed from racks, and frequently fatal to each other; for in the rutting season they will fight with such fury as to occasion the death of one of the combatants. Few of the Spanish Merinoes are found hornless; when such are found, if they are equally perfect in other respects, they should be preferred. I have one of this description, of fine size and figure; but the materials that should have formed his horns appear to have been transferred to his hoofs, which grow out to the length of six or eight inches, and must be cut at least twice in a season, or they render him lame; the cutting, however, is attended with little difficulty, and no danger.

Most people observe the season of shearing for docking, castrating, and marking their lambs. I believe the docking contributes to cleanliness, and therefore have adopted it. In England, however, it is seldom practised, but very commonly in Spain. The castration is performed in various ways; some prefer cutting out the testicles, while others tie the scrotum so as to stop the circulation, and after four or five days, when the parts are dead, they cut it off just below the string, and tar the wound. This is said to be the best mode, where the castration is to be performed on an aged sheep. But great attention should be given to the weather: I lost six out of seven last year by having it performed in warm weather on sheep of only three months old. When the lambs are young, cutting is the easiest and best mode; and if the season is advanced when they drop, they may be safely cut at ten days old; indeed, the earlier it is done the finer will be the wool and the flesh; but if rain or cold weather succeeds before they are cured, they should be housed, as I have known them to die for the want of this precaution. In Spain it is usual, instead of either of these operations, to twist the testicles within the scrotum, so as to knot the cord; in which case they decay gradually, without injuring the sheep.

When it is desirous to confine the breeders in a flock to a less number than that of the ewes produced, the ewe lambs may be splayed; in which case they give more wool, fatten better, and are said to afford finer meat. This operation cannot be performed conveniently before the lambs are six weeks old, because the ovaria is at an earlier period too small to be easily distinguished. They are said to suffer so little by this operation, as not to feel any inconvenience from it after the first day.

The time for weaning the lambs depends upon various circumstances. If the parent ewe is broken mouthed, or so faulty in wool or in shape as to render it desirable to get rid of her, the lamb must be weaned early, so as to admit of her being fatted in season; if she is admitted to the ram as soon as she is disposed to take him, the earlier she will fat. If the object is to render the lambs as large as possible, and they are of such a stock as to make the ewes of comparatively less value, it will be best to let the lambs run with them till they wean themselves, because they undoubtedly grow the more rapidly for it. This mode I would therefore recommend when a Merino flock is to be engrafted upon a common one. But if the ewes are valuable, it certainly will be best to wean the lambs so early as to give the ewes some respite before they take the ram again; and indeed, if early lambs are preferred, early weaning is absolutely necessary, as the ewe will seldom take the ram while exhausted by nursing. In Spain they leave the lambs with the ewes till they wean themselves. In France, and generally in England, they are weaned at three or four months old. In order to prevent the lambs from falling off when they are weaned, they should be put into a piece of young tender grass, with an old quiet ewe or wether to direct their movements; they should also be out of sight and hearing of their mothers, that they may the sooner forget each other. If the keeping them apart should be inconvenient, they may be brought together at the end of a fortnight. Some attention should be paid to the ewes for the first week, in order to prevent their suffering by a too great flow of milk, which should be taken from them every day or two; and perhaps it would be best, till their milk was dried up, to keep them in scanty pastures.

It was the opinion of the Romans that the first lamb from a ewe was generally weak and pot-bellied; they separated such from their flocks, and fatted them off. I believe the opinion well founded, but I think it arises from the young ewes seldom having so much milk.

or being so careful of their lambs as the older ones. If the lambs come early, it will be necessary to wean the forward males before the first of August, particularly if the ewes are in high order, or if some among them have lost their lambs early, as they may otherwise impregnate the ewes sooner than is proper. It is a very common practice in Europe to shear the lambs, though it is seldom done here; and yet I think it more adapted to our climate than to that of northern Europe. The heat of our summers renders the wool very burdensome to the lambs; and as our autumns are generally fine and dry, there is sufficient time for the wool to grow so much as to protect them during the winter. Lamb's wool also sells much higher here for hatters' use than in Europe, so as to render the shearing more a point of profit. After the lamb is shorn, he should be washed and perfectly freed from the tick. Though I do not wash my sheep before shearing, I always have them washed, after they are shorn, once or twice during the hottest weather, and think that the practice is useful in freeing them from tick and preventing the scab. My lambs will drop very early this year; I contemplate shearing all those not intended for sale, and washing them not only in running water, but with soap.

I propose to make soap, for the purpose, of turpentine instead of grease, and to mix a weak decoction of tobacco with the water in which they are washed. This I think will not only free them in the first instance from tick and lice, but keep the tick fly from assaulting them as long as the scent of the tobacco or the turpentine remains. Tar, as a cheaper material, may be used instead of turpentine; neither of them will injure the wool when mixed with a just proportion of alkali, and diluted when used.

The next care of the attentive shepherd is to examine his flock frequently, in order to know whether they are in health, and to remove such as he may find distempered; for almost all the diseases of sheep are contagious, and a whole flock may be lost by the negligence of a few days. Happily, that long catalogue of disorders which prevail in other parts of the world, is reduced in our country to two—the scab, and the staggers, or dizziness. The scab is a cutaneous disorder; it is discoverable from the sheep's frequently rubbing himself, biting and pulling his wool; and, when it has made some progress, from the wool's rising or falling off, at first generally on the back. Fine-woolled sheep, though not more subject to this complaint than others, are, it is said, more difficult to cure, both because of the closeness of their coats and the delicate texture of their skins; I have, however, found very little trouble with it, though it has appeared several times among individuals of my flock. The care that I take to have the flock drawn up and examined at least once a month, has prevented its spreading when it appeared, as the infected sheep were immediately separated, and means used for the cure, which has never in one instance failed to be effectual after having been three or four times applied. For a particular account of my remedy, and several others that are applied to this disorder, I shall refer to the Appendix, in which I shall enumerate the diseases of sheep, with the most approved remedies.

The staggers, or dizziness, which is also known by various other names, has occurred in three instances in my flock, and always attacked lambs under one year old; and, indeed, I believe it is confined solely to lambs. They were taken very suddenly, and without any previous symptoms, by a species of convulsion, in which the neck was twisted to one side; they lost the use of their legs; when raised they would attempt to follow the flock, but turned round and fell; in a few days they were incapable even of standing, of moving their heads or any of their limbs. As they were very valuable sheep, I paid particular attention to them; grass and grain were given them, which they would readily eat, though they could not move any part but their jaws. In this state they lay a week without motion, except of their eyes and mouth when food was given them; they then so far recovered as to be able to stand when they were supported, in which posture their food was given them, but would fall down when the support was withdrawn. In about ten days they could stand without support, but fell when they attempted to walk; their motion being rather a convulsive run than a walk. At intervals they would get better, and be able to walk for some time, but they were always found laying in some part of the field as if they were dead. Observing that the vivacity of their eyes was not altered, I directed that the attention in feeding and supporting them should not be remitted, and in the course of about six weeks they so far recovered as to join the flock; one of them, however, a young ram, received a blow in his weak state from a stronger one, that killed him; the other two recovered, but very slowly; and even at the end of eight months they bore evident marks of their complaint. This disorder is found, upon dissection, to be owing to a bag containing water within the skull, which presses upon the brain. It is generally considered as incurable, though it is said by others that it may sometimes be remedied by trepanning: a soft place on the head indicates the situation of the bag, which, if taken out whole, will remove the disorder; others pass a sharp wire up the nostril into the brain, and perforate the bag: the suppuration that this occasions effects the cure; five out of six, however, die under this operation, and it may, therefore, be justly considered as incurable by the doctor, but not, as I have shown, by the nurse. Nature will effect the cure, if care is taken to feed and tend the patient while she is operating her very reluctant and tardy cure; but a sheep must be extremely valuable to pay for three months constant attention. I should add, that I bled the lambs I mentioned, and gave them a doze of train oil; but I have no reason to think that either of these had any agency in the cure.

The purging which sheep are subject to in the spring of the year, and which arises from their change of food, I do not consider as a disease of any consequence, and except this and the staggers I know of none that prevails in our flocks when properly nourished. When they are ill kept, they sometimes take colds and discharge a mucus from the nose. Good feeding and pine boughs, or tar and salt, administered in the manner I have mentioned, will cure this complaint.

It is frequently asked, what quantity of food, either dry or green, is necessary for a given number of sheep? and the inquiry is not a mere matter of curiosity, but its answer very important to the farmer, as it enables him to adapt his stock to his means of support. The British writers are not so accurate on this subject as one could wish, and as they generally are in whatever relates to rural economy. This is owing to the manner of feeding their sheep for the most part on turneps eaten on the ground, on old grass fields, and only occasionally on hay. Happily, however, this interesting question is answered by Daubenton, a celebrated French agriculturalist, in such a manner as to leave me nothing to do but to transcribe his work. "I confined, "in a small space, two sheep, about twenty "inches high (the height of most woolled ani-"mals in France). By way of experiment, I "caused the sheep to be fed, during eight days, "solely upon grass newly cut, and weighed be-"fore placed in their rack. Care was taken to "pick up, and place in it back again, all that the "sheep let fall, and to weigh that which they

" would not eat in consequence of its being too " tough, or because it possessed some bad qua- " lity. From this trial, frequently repeated, it " appeared that a sheep of the middle stature " eats about eight pounds of grass in a day. " The same experiments, conducted with the " same preciseness, in regard to the fodders of " hay or straw, have proved, that a sheep of " middling height likewise eats daily two " pounds of hay, or two pounds and a half " of straw.

" In order to ascertain how many pounds of " grass go to one pound of hay, I caused the " grass to be weighed as soon as cut; it was then " spread on cloths exposed to the sun, so that " none might be lost, though at the same time " well dried. Being thus converted into hay, " I found its weight reduced to one-fourth; " eight pounds of grass had only given two " pounds of hay.

" Agriculturalists know how many cart loads, " or trusses, a field can produce; consequently " they may judge how many sheep it can " maintain in hay or in grass. They have a " rule then for proportioning the number of " their sheep to the quantity of pasture and " fodder they can supply them with.

" Having determined the quantity of solid

"food essential to the good regimen of the "woolled kind, I made other experiments upon "these animals, in order to know at what time "they should drink.

"It is well known that they seldom drink "when they feed upon fresh grass, but stand in "want of water when fed on dry meat. Dif-"ferent opinions are pursued as to the proper "time for watering them. In some countries "they are taken to water once or twice every "day; in others not for one, two, three, four, "or even five days. By the following experi "ments, I have endeavoured to ascertain which "of all these regimens, so different from each "other, is entitled to preference.

"I shut up in a stable, in the depth of win-"ter, a small flock, of which all the sheep were "marked with a number. They were kept, "night and day, without being suffered to quit "it, and fed with a mixture of straw and of hay, "without any other aliment. Each day a shep-"herd carried in his arms, successively, some "sheep out of the stable, to let them drink in "my presence, out of a vessel guaged at dif-"ferent heights, and then took them back into "the stable, when they had either drank or "refused to drink.

"By this method I knew how much water

" the sheep had taken, when presented with it " once, twice or thrice each day, or only once " in two, three, four, or five days.

" Most of the sheep in this little flock passed " a month in the stable without drinking: their " appetite was always the same, and they expe- " rienced no other inconvenience than that of " thirst, of which they gave evident proofs by " running to lick the moist lips of those carried " back to the stable on return from drinking.

" The result of these experiments, which I " cannot here detail, led me to conclude, that " sheep, with no other nourishment than that " of dry hay, and within reach of water, could " pass days without drinking; but they would " take a greater quantity of water the following " day than if they had drank the evening be- " fore: this quantity increases to a certain de- " gree if they have been deprived of water for " many days together. They are then tor- " mented with thirst, for they are eager to get " a drop of water; if they could find it in " abundance, they would drink too plentifully " for their temperament, subject as they are to " effusions of serosity, which produce mortal " hydatides in the brain, and the rot, a disease " no less fatal.

" The best plan is to drive the flock every

" day to the pond, and to make it pass slowly, " without stopping there: by this method it " will be found that the sheep who really want " to drink will be the only ones who will drink.

" In countries where water is scarce, it fre- " quently happens that the pond, if far distant, " and the flock cannot be driven to it without " being fatigued; in this case they may pass " many days without drinking; but when fed " only upon dry meat, it must not be delayed " too long.

" This aliment differs much from fresh grass, " in consequence of the loss of moisture by dry- " ing; yet sheep take daily the same quantity " of solid food, whether in grass or in hay. In " the experiments before mentioned, I found " their appetite perfectly equal, for they eat " eight pounds of grass, or two corresponding " pounds of hay, which I found to be the pro- " duct of eight pounds of grass. The evapo- " ration which is carried on during the making " of the hay, takes off three-fourths of the sub- " stance of grass in fluid particles; thus the " sheep which eats two pounds of hay is de- " prived of six pounds of liquid aliment, which " it would have had by eating eight pounds of " grass. It supplies a part of this deficiency " by drinking about three pounds of water "when fed upon hay; but this water is not in "sufficient quantity, and possesses not the same "quality as the liquid of the grass evaporated "in drying.

"There can be no doubt that this difference "in regimen is productive of bad effects. I "shall mention some proofs of it, which are "indeed too evident and too frequent.

"In countries where the snow remains upon "the ground for one or two months, the cattle "are reduced to dry fodder so long as it lasts; "then the weaker sheep, and chiefly the lambs, "the sheep of the second year, the pregnant "ewes, and those in milk, languish and drop "off. Shepherds denote this miserable state by "saying, they melt their fat: they certainly "grow very lean, and fall off in great num-"bers.

"I have often reflected upon the cause of "this evil, and the means of preventing it. "After having prosecuted every inquiry I could "think of, it appeared to me to arise solely "from a change of diet too suddenly effected. "In one day the sheep are reduced from eight "pounds of grass to about two pounds of dry "fodder and three pounds of water. They "are thus deprived, therefore, all at once, of "three-eighths of their wonted nourishment,

" and these three-eighths composed the half of " the fluid part of it.

" According to my experience of the quan- " tity of water taken by sheep, it appears that " their drink can only supply one-half of the " liquid which grass contains more than hay. " It would be dangerous to excite them to drink " a greater quantity of water, because they are " very subject to infiltrations. We must, there- " fore, endeavour to supply them with at least " a small quantity of fresh food every day, in " order to correct the bad effects resulting from " dry meat.

" The most sensible of these bad effects ap- " pears in the third stomach, composed in the " interior of a great number of membraneous " folds, detached one from another, although " it is only from eight to ten inches in circum- " ference when filled with air. During rumi- " nation, the food passes from the throat into " this third stomach, and spreads amongst all " these folds. I have there found it very fre- " quently parched, and almost withered, in " many sheep which I have dissected.

" This aliment, after having been ruminated, " receives, in the third stomach of the sheep, " and of other animals that chew the cud, a " preparation for digestion, which latter takes

" place only in the fourth stomach. The ali-
" ment is dry in the third stomach, not only
" when the animal is fed solely upon dry meat,
" which has not furnished sufficient liquid, but
" also when attacked by some disease causing
" too great heat, and consequently too great
" evaporation of the liquids necessary to diges-
" tion. In these cases, bad digestion, and the
" evils attending it, may be prevented by giving
" some green food at least once a day.

" At all times when the ground is not covered
" with snow, sheep find upon it sufficient fresh
" food to render it unnecessary to give them
" any in the rack with their dry meat, in a bad
" season. I have often stopped in the midst of
" a flock, in fields half covered with snow,
" where no grass whatever was to be seen; the
" sheep, however, having their eyes near the
" ground, perceived the points of some leaves,
" and scratched with their feet to find more of
" the plant; they then seized it with their teeth,
" and sometimes pulled up the roots along with
" the leaves. But when the snow entirely co-
" vers the ground to a certain thickness, there
" is no other recourse than in the plants which
" are high enough to enable the sheep easily
" to remove the snow which covers them.

" There are many kinds of cabbages, such

" as the fringed cabbage, which are very tall; " they resist the frost, and their leaves contain " much juice. These form an indifferent arti- " cle of food for sheep in times when they are " not reduced to dry meat; but, if confined to " this aliment, a few of the leaves of these plants " will be found sufficient to obviate its prejudi- " cial effects.

" It is difficult to have a quantity of these " cabbages sufficient for numerous flocks; they " require to be sown, transplanted, and watered " for many days; and this culture must be re- " peated every year, which is too tedious and " expensive for the husbandman. Whatever " advantage may attend the use of cabbages as " a diet for sheep, I would not recommend this " plant as fodder, had I not met with a species " of cabbage which may be reared without sow- " ing, without transplanting, or watering. It " is equally unknown to the naturalist and to " the agriculturalist. Like the fringed cabbage, " it resists the frosts; and, for cattle, is preferable " to it, being very easily cultivated. It may be " propagated by cutting; it is only necessary " to slip off its lateral branches, which are nu- " merous, and plant them in the earth, to have, " in a short time, new plants over the whole " extent of a well cultivated field. The leaves

" are less than those of other cabbages, but the " juice they contain is as abundant; they are " equally good food for the shepherd as well " as his flock. Some handfuls of these leaves " given to a sheep, will correct the bad effects " of dry food.

" The regimen of sheep is one of the im- " portant branches of veterinary medicine. This " science is to be established only by well- " founded experience, with observation and " experiment frequently repeated on these ani- " mals. An intimate acquaintance with them " in their natural state, is necessary before at- " tempting to cure their diseases."

CHAPTER III.

MERINO SHEEP.

ONE of the principal objects of this Essay being to impress upon my fellow citizens the importance of cultivating this most valuable breed of sheep, I propose to devote this chapter to them, in addition to what I have already offered.

One of the first ideas that strikes the farmer is, that his sheep may degenerate, and that if the quality of their wool should change, he would have put himself to great expense to change a sheep of better size and form for one which he imagines to be inferior in both; and he is strengthened in this opinion by having observed, that most of the British sheep that have from time to time been brought here, have degenerated. This I confess very generally to have happened, but I deny that any inference injurious to the Merino breed can be drawn from it. The British sheep here alluded to are the long-woolled, for no others were thought better than our own. This race of sheep can only be advantageously maintained on rich and luxuriant pastures, and an ample supply of succulent food during the winter. Experience has taught us that rich pastures will add to the length and quality of wool on our native sheep, and that bad keeping will diminish it. Without attention to this circumstance, the long-woolled sheep have been transferred from the fens and marshes of England and Holland to our dry, short, sweet pastures; from which it was expected that, labouring under a thick coat of long wool, and contending with our summer sun, they should be able to fill their large carcases. Not having pastures adapted to their size and their habits, they could not subsist but by gradually accommodating themselves to ours. This necessarily occasioned a diminution, first in the quality of the wool, and next in the size of their descendants; besides, that it was very rare to obtain the full bred sheep, both rams and ewes, and to preserve them unmixed. If the rams bred with our ewes, their progeny would soon be reduced to the size of the ewes; directly, because of the mixture, and, indirectly, from the ewes not being able to afford nourishment to a larger stock than nature designed her to support, without the most uncommon care in feeding her while she gave milk.

It is always for this reason very injudicious to breed from the females of any stock of a race inferior in size to that of the sire, since they will in such case necessarily degenerate. The reverse will take place where the ewes are larger than the stock from which the rams spring. The lambs being abundantly nourished, will, by degrees, attain the size of the dam, while they preserve the other peculiarities of the sire. It is by attention to this circumstance that I have already greatly improved my Merino stock in size and beauty, when I have bred them in the fourth generation from the finest ewes of the country; and where I bred from imported ewes I have attained the same object, by affording them a plentiful supply of food while they nourished their young. As these ewes were themselves of the largest stock of Merinoes, I have gradually added to the size of their progeny; and I have now full-bred Merinoes at Clermont that are larger than the common sheep of the country; and my half and three-quarter-breed wethers are, when stripped of their coats, larger and much handsomer than most of our native flocks. When the fleeces are on, there is some deception in judging of long-woolled wethers, as they seem larger, and their defects are concealed by their covering; whereas the short, close wool of the Merino shows his shape precisely.

But to return to the question of the degeneration of the Merino sheep. So far as a scarcity of food may, as I have said, operate a change for the worse in sheep, it cannot apply to the Merino when introduced into our country; because, not requiring better pastures than our own sheep, there is no reason for the change of size, at least such change as the wool of those sheep that have been introduced from Britain has undergone: this was a change in the quantity rather than in the quality. When a sheep diminished in size, it would have been a very unwise provision of nature to have suffered him to carry the same quantity of wool which he bore upon a larger and a stronger carcass; his wool, therefore, diminished in length in the same manner that his carcass did in size; but the quality of the wool remained the same, or, if any thing, changed for the better. So if the large and improved breed of Merinoes were kept upon very scanty pastures, they would diminish in size, and carry shorter fleeces; but those fleeces, even under the worst keeping, would still retain all their original properties. We are often told of the influence of climate in effecting changes: that it operates I can believe, but I also believe that it operates very slowly, and that until experience has determined the fact, it is impossible to say whether that operation will be for the better or for the worse. For my own part, I believe that the change in the Merino sheep brought into any northern country, provided they are plentifully fed, will be for the better, and particularly when brought into this State, where the pastures are good, the air and waters pure, the winters cold, and the summer range furnished with shade. I should have presumed this in reasoning *a priori*, and I have found my theory confirmed by actual experiment.

I am now to mention a circumstance on which I ground my reasoning, which may appear fanciful to those who have not attended to the proofs of the improvement of Merino sheep in high latitudes. The Merino differs more essentially from every other kind of sheep than the Spanial does from the Mastiff, and yet no one has seen any change in either of those species of dogs in a course of generations, or in any climate, except by intermixture of the breeds. I say the Merino differs essentially from all other sheep, and even from all other quadrupeds of which we have any knowledge, as an annual does from a perennial plant. All quadrupeds change their coats every year, and indeed generally twice a year: the Merino sheep never changes his coat; on the contrary, it will continue to grow from year to year, and at the end of the third year the fleece will yield a three years crop, with little or no diminution. This experiment has been tried in France, in Switzerland, and in England, for the course of three years successively, and always with the same result. The wool of this sheep then resembles in its duration human hair, and may probably be subject to the same physical laws. Human hair is affected by the tissue of the skin through which it passes. In warm climates the hair of man is generally black and coarse; in cold ones we find flaxen, yellow, and various shades of brown, to be the prevalent colours; and even where the hair takes a deeper shade, it is finer than the lank black hair of the south. May not this be owing in some sort to the skin being more braced in one and more lax in the other? and will it not produce the same effect upon the wool of an animal whose fleece is perennial, particularly if the food and air invigorate at the very time that the climate braces the fibres? It is said that the wool of the common sheep is sometimes coarser, as he is either well or ill fed. This may happen if he is either sickly or in full health, or if the weather is more or less cold when the young wool protrudes through the skin: if in that state it is compressed, it will be fine; if it finds an easy passage, it will be coarse; and as the wool of common sheep is an annual production, it may frequently vary. But the fleece which never falls off must be subject to very few changes; it may be longer or shorter, but the root being the same, it will probably be liable to no changes but such as arise from the greater or less compression of the skin through which it passes. Cold then will have a tendency to render the wool fine; heat and moisture to make it coarse. The marten, the grey-squirrel, the common fox, &c. have much finer fur in Siberia and Hudson's Bay, than they have in Virginia or Pennsylvania, and yet they are exactly the same animal. It is true, the men of very high latitudes have similar hair to those near the line, and probably this is owing to the same cause: in summer they are exposed to the continued rays of the sun, without the intervention of night, which must greatly relax them: their winter is a continued night, in which the children at least are confined to a smoky hut; their diet is slender and relaxing; and the general habit of covering their heads and greasing their bodies must necessarily tend to unbrace the skin and give an easy passage to the hair. We find an exact analogy between the effect of climate upon the covering of sheep and that of other quadrupeds. The sheep under the line are hairy; as you go north they become woolly, and farther north the wool is finest; the best wool in Germany is that of Saxony. The moist climate of England and Ireland produces long and coarse wool. It is true that fine wool is also found in Persia, and in Cassimere and Thibet, but this is only in the very cold and mountainous parts of those countries. The sheep of Siberia are coarse-haired, but they have below that hair a coat of extremely fine wool; they are the Mouflon, or Argali, almost in their native state, in which man has taken little pains to cultivate the wool at the expense of the hair, but permitted them to grow together; and indeed in that state it is best adapted to the wants of the inhabitants, who know not the use of the loom, but wear the skin of the sheep, in which case the hair is as useful as the wool; for it protects them, as it did its original owner, against rain and snow, which would penetrate the wool were it not covered by a surtout of hair: it is then probable that the Merino sheep does not owe its peculiar excellence to the climate of Spain, or to the mode of treatment.

Spain, as I have said, contains a great number of long-woolled sheep, in every respect different from the Merino: the climate has had no effect in meliorating their fleeces; the migration does not contribute to it. They have in various parts of Spain, and particularly in Estramadura, Merinoes that never migrate, and whose wool is not inferior to that of the migrating sheep; and they have both in France and Italy migrating sheep whose wool is not fine.

When nature forms a change in any species of plants or animals, it does so very slowly, and always in such a way as better to adapt them to the climate in which they are to be naturalized. Thus, some plants which are perennial in warm climates, both root and branch are annuals in colder ones; or while the roots of others survive the winter, their stems are annually renewed. The same plant will form a tree in one climate and a shrub in another. This I have myself witnessed in the fig, which I have seen of the size of a bearing apple-tree, while a little more north it was a shrub of very moderate size. If then the fur of quadrupeds and the hair of man are finer in high than in low latitudes, why, if the climate effects any change in the Merinoes, should it not be for the better? My own experience has not been so great as to permit me to build much upon it, since my sheep were only introduced in 1802; but as far as it goes, it leads me to believe in the amelioration of the sheep, either from the effects of climate, or from attention. The original stock were chosen with very peculiar care in France, after the most careful examination of their descendants; they have improved in size, beauty of form, and quantity and quality of the fleece. The two first improvements are too obvious to admit of the least doubt; the last requires so nice a discrimination as to make the decision more difficult in all but one instance, where the difference is so striking as to be evident to every observer. I refer to a ram lamb of the last spring, who is out of an imported ewe, while his sire (who is also by the same dam) was bred upon my farm. This lamb is of the most uncommon size and beauty; his fleece, compared with that of any other of my improved sheep, or with any sample that I have been able to obtain of others, is indisputably much finer, and, at the same time, so long and abundant, that I have little doubt of his yielding at least eight pounds of wool the first shearing.* I imported the summer before last a very fine ram, whose fleece has, by

* He has been shorn since this went to the press, and gave nine pounds six ounces of wool.

the best judges, been pronounced superior to any they had examined; yet his wool is certainly not better than that of the lamb I speak of; and this is the more extraordinary, as the Merino lamb's fleece is never so fine as his subsequent growth.

The account I have already given of the flock at Rambouillet shows, that instead of degenerating they have greatly improved in the fineness of their fleeces. Dr. Parry, who has lately written a treatise on the Merino sheep in England, acknowledges that the wool of the Rambouillet flock is finer than that imported from Spain, and speaks of this flock in the highest terms of admiration: he also adds, that the flock of Lord Somervile, and of his Britannic Majesty, as well as his own flock of Merinoes in the fourth generation (fifteen-sixteenths), are finer than the wool brought from Spain to England, and proves it by showing that it requires two pounds of imported wool to make one yard of the finest British broadcloth, and that he has made from his Merinoes upwards of twenty-six and a half yards from forty-two pounds. This is something more than one pound nine ounces to the yard. If I was to determine the fineness of my flock by the same rule, I should exceed both, since the same quantity of cloth was made at Clermont by common country spinners and weavers from one pound four ounces of Clermont Merino wool; and thirty-two and a half yards of twenty-five and a half inches wide, were made in Mr. Edward P. Livingston's family from sixteen and three-fourths pounds of wool.

In the year 1723 Merino sheep were carried to Sweden, where they have greatly multiplied, and retained their original purity. If long cold winters, and even bad keeping, would change them for the worse, they would have experienced that change in upwards of eighty years, during which they have so greatly multiplied as to have stopped the importation of Spanish wool into Sweden. Indeed, the experience of a number of other nations has put it out of doubt, that the Merino sheep do not degenerate by being carried to a cold climate. This fact being once established, what is to stop their progress in our own country? I have already shown how the most indigent cultivator may, in the course of a few years, convert his common sheep into Merinoes, not only without expense, but with profit. If he fears that they are more delicate, and require more care than our common sheep, I can assure him, from my own experience, that though like all others they will be the better for being well kept, yet they will not suffer more from neglect; their thick and close fleeces fit them for bearing cold, and they will in every mixed flock be found among the most thrifty in the severest weather. The objection to their size I have shown to be ill founded, if he draw his stock from the improved Merino; and even if he begins with one of the small race, he will, in some years, by breeding out of good ewes, advance gradually to the size of the dams. Nor let him be under the least apprehension that he will not in this way have as fine wool as if he bred out of full-blood ewes. It is now so well established as not even to admit of the smallest doubt, that a Merino in the fourth generation, from even the worst woolled ewes, is in every respect equal to the stock of the sire. No difference is now made in Europe in the choice of a ram, whether he is a full-bred or fifteen-sixteenths. Indeed, Dr. Parry maintains, from his own experience, that they are superior to full-blood rams. He says that the wool of his flock (which consists of sheep in the fourth generation from the Ryeland ewes) was injured when he put a fine full-blood Spanish ram to it; and asserts, that any person beginning a stock with an imported ram, will be eight years behind one that begins with a fifteen-sixteenths of the Ryeland Merino; and I can easily believe that there is some justice in the remark, since I cannot conceive that one-sixteenth of common blood, which will only be one-thirty-second in the offspring, can make any difference in the fleece; whereas considerable difference may be occasioned in the beauty and vigour of the flock by the ram having been bred, for four generations, from ewes of the country, assimilated to the climate and to the manner of keeping.

As I have mentioned the Ryeland ewes as the basis on which Dr. Parry formed his stock, it will be proper to give some description of them, otherwise it might be thought that they possessed some peculiar excellencies not to be found in our sheep. An account of them is inserted in the Annals of Agriculture, vol. xx. p. 15. They are short-woolled sheep, yielding fleeces of from one and a half to two pounds. The best of these fleeces sell at two shillings and six-pence sterling the pound, without the breachings; they weigh, when fat, fourteen pounds a quarter. From this description it is pretty clear that they are not better than our short-woolled New-England sheep, and yield less wool. It has always appeared to me, that our native stock has been injured in this State, and in many other places, by crossing them with long-woolled sheep; and upon this idea I have founded the recommendation I have offered of short-woolled sheep, as forming the best stock whereon to graft the Merino breed, provided the ewes are large and well made.

Having mentioned Dr. Parry's concurrence with the French agriculturalists in the opinion that the breed is completely changed in the fourth generation, I should add, that he mentions *one* instance in which it was not. This was of a Merino bred on a Cape ewe. But I think this proves nothing, because a Cape ewe has not wool, but hair; and because he had no means to ascertain that the sample shown him had really undergone no other cross. The French agriculturalists say, that however coarse the fleece of the parent ewe may have been, the progeny in the fourth generation will not show it: and, indeed, I have seen, and deposited with the Society of Useful Arts, samples of wool from sheep of every description that could be procured in France crossed by Merinoes, and can discern no difference between those in the fourth degree and the original stock. It follows then, that any farmer may, in the space of six or seven years, convert his common flock into Merinoes, with this great advantage, that during the whole of his progress he is annually adding to the value of his fleeces, and selling off old sheep instead of lambs, thus reimbursing himself for the expense of his ram, which is the only extra expense he has sustained; and he is also parting with a number of male lambs at a higher price than he was accustomed to receive for those of his old stock. The wool of a common flock barely pays the keeping; their only profit arises from the sale of sheep and lambs, which, supposing the flock to consist of fifty ewes and fifty wethers and rams, and that thirty-five are sold off yearly, which is as many as can be calculated upon with those necessary to keep up the stock, the clear profit will be seventy dollars upon one hundred sheep. An half-blood flock will bring, in the increase of quantity and value of the fleece, one dollar and more upon each sheep,* even counting the sales of lambs at the rate of common sheep. The second year then, the purchaser of a ram will receive one hundred and seventy dollars profit, instead of seventy. When the flocks are three-fourths breed, his wool will rise to eighty-one cents in the pound. (I state the lowest rate, mine of that grade sells at one dollar). This will give him a

* The difference of profit between the half-breed and the common sheep, at my last shearing, was two dollars and six cents per head.

clear profit of one dollar and fifty cents per head beyond the value of his old fleece, or one hundred and fifty dollars added to the price of sheep sold at seventy, bringing his profit to two hundred and twenty dollars clear of all expense. When his flock consists of seven-eighths breed sheep, his wool will rise to one dollar and twenty-five cents the pound; I sell mine at one dollar and fifty cents. Supposing the fleeces of his ewes and wethers, taken together, to weigh three and a half pounds, his flock will bring him, after deducting all expenses, which I rate at one dollar and fifty cents per head, two dollars and seventy-five cents each, exclusive of lambs; that is, two hundred and seventy-five dollars; which, added to the sheep sold, seventy dollars, makes a clear profit of three hundred and forty-five dollars annually. When his flock are full-bred, he will receive two dollars per pound for his wool, which, at three and a half pounds the fleece,* will give him seven dollars per head, or, deducting the keeping, five and a half dollars; that is, five hundred and fifty dollars, added to sheep sold, seventy, making an annual profit of six hundred and twenty dollars instead of seventy, which his common sheep

* Mine averaged the last shearing upwards of five pounds the ewe's fleece.

would have brought him. In this I have stated nothing for the increased value of the lambs sold, lest it should be said that no sale may offer for them. This, however, is an error, in a country so rapidly increasing as ours, and which does not grow one-fifth of the wool necessary for its own consumption; and when all the stock of sheep will be converted by intelligent farmers into Merinoes, there will be a demand for lambs for at least twenty years, at an advanced price; so that I have no hesitation in saying, that the profit upon the lambs will be more than equal to that of the wool. To state the account fairly then, the annual profit should be doubled. Provided the farmer sets out with the best stock, and takes care to breed only from good ewes, he will find demand for any number he may wish to part with.

I have already anticipated what it was necessary to say as to the choice of the ram, and the manner of forming and keeping a flock. It may, however, be said, that when this breed is more diffused, the price of the wool will fall. I am not of this opinion, because, besides our own, there will be a foreign demand. This wool now sells in England at seven shillings and three-pence sterling, and is constantly rising. But admit that it should fall, it is certain that common wool will fall much more rapidly; because, when habituated to fine soft cloth, few will wear the harsh, hard, heavy clothing we are now content with, particularly if fine wool is reduced in price. The relative difference between the Merino and common sheep will not change; if the Merino wool brings less, the common wool will not bring enough to pay for the keeping of the sheep.

So much respect is due to the opinion of Mr. Custis, who has laboured with great zeal and success in the improvement of sheep, that it cannot but be proper here to state and consider his reflections on the Merino breed of sheep, contained in his very valuable publication. He thinks that the Merino breed will not be generally extended, because of the high price at which the rams are held; one hundred dollars being, as he supposes, the price of a ram, which was that at which they were sold when he wrote, but they have since risen to one hundred and fifty dollars. The reverse I conceive to be the case; the high value at which they are rated will continue in two ways to extend the breed. First, by yielding a great profit to the breeder; and, next, by introducing more from abroad. If a farmer believes he can sell his half-blood rams or ewes at twelve dollars, *their present price,* he will more readily purchase a ram at one hundred and fifty dollars, than he would have bought him at one hundred, if he could only sell his lambs at two dollars. One ram will bring him fifty lambs; this, at twelve dollars, is six hundred dollars: with a ram at ten dollars he could have fifty lambs, worth two dollars, which is one hundred dollars; deduct the price of the ram, and in one case he gains ninety, and in the other four hundred and fifty dollars the first year; the second year he gains in the one case one hundred, and in the other six hundred dollars, the expense of keeping being the same in both stocks. It was not till the price of rams rose very high that any important improvement was made in British sheep, and this is precisely the case in this State. All the full-bred rams of the Clermont stock were bespoke before the first of January, at one hundred and fifty dollars; and one thousand dollars has been refused for the ram lamb of ten months old that I have before mentioned, and two hundred for his brother, dropped at Christmas, and only three weeks old when the offer was made by an enlightened farmer of Massachusetts. What is all farming but an advance made with a view to future profit? No man refuses to sow wheat because the seed is dearer than rye. A rich

Virginia farmer, who puts in one hundred acres of wheat, cannot estimate the ploughing, harrowing, seed, and harvesting, at less than five dollars and fifty cents per acre; his returns, if I rightly remember the information I received from our departed hero, Washington, will fall short of seven bushels to the acre; thrashing and carrying to market will amount to about fifty cents more; so that upon a capital of five hundred dollars he seldom receives two hundred, taking wheat at its average price for the ten years last past. This falls greatly short of the profit upon the amount of two Merino rams put to one hundred ewes, if the lambs were sold at the rate I mention; but putting the lambs out of the question, and supposing the profit to be made only upon the fleeces, then a farmer who had a fine flock of sixty ewes, averaging three and a half pounds of wool, worth thirty-seven and a half cents, that is, one dollar thirty-one and a quarter cents each, (the lambs paying the expense of keeping,) would, by putting the same ewes to Merino rams of the improved breed, gain a stock of lambs which would the first year give him fleeces weighing four pounds, worth seventy-five cents the pound, and a ready market: thus he would gain upon sixty ewes half the price of his ram the first year, and progressively more every year as he reformed his stock.

The second objection to the Merino is the high price of his wool, which Mr. Custis supposes can only be adapted to the use of the rich, while the low price of the common wool fits it for general consumption. If Merino wool can be raised as cheap as that from common sheep, it comes at least as cheap to the grower, and therefore he may wear a coat of fine wool at no greater expense than one of coarse cloth; and there can be no sort of doubt that if it is manufactured exactly as the other, the coarse cloth made from fine wool will outlast two made from harsher materials, and, at the same time, be warmer. If blankets and flannels are a domestic manufacture, both these articles will come as cheap to the growers of the wool as if made from the long-woolled sheep, and certainly will be infinitely warmer and lighter. If a man's land is such as to bring him a good crop of wheat, he certainly will not sow rye or buckwheat, but will prefer wheaten bread for his family; or, if he is an economist, he will sell his wheat and buy rye. Is it not exactly the same with Merino wool? If he is in easy circumstances, he will manufacture it, and sell the excess beyond what he wants for his own consumption, at such a price as will pay his weaver, his dyer, his dresser of cloth, his tailor, and perhaps as much more as to pay for the keeping of his sheep: whereas, if he raises common sheep, unless he keeps much larger flocks than are necessary for his own use, he has nothing to pay these expenses; nor indeed, in the northern States, whatever be the size of his flock, can he sell any thing, since the fleece hardly pays the keeping. Ten Merino sheep beyond those whose fleeces he employs, will give him thirty-five pounds of wool, which will sell at seventy dollars, the present price of Merino wool being two dollars the pound. This will not only pay for that part of the manufacture of cloth which is done out of the family, but will leave him an excess for other purposes. There are few farmers that cannot spare the wool of ten sheep; but if these were even very good common sheep, their fleeces, at three pounds, would amount to no more than fifteen dollars; so that he must draw upon some other fund to pay the tradesmen employed in clothing his family. Which stock of sheep then is best adapted alike to the poor and the rich? certainly that which, after furnishing the material, pays for making the cloth by the high price at which a small excess is sold. Mr.

Custis, however, presumes that much larger fleeces are obtained from other sheep, particularly from the Smith's Island and the Arlington. It is possible that Mr. Custis has drawn this inference from not having seen the improved Merino breed, and perhaps in this view his deduction may be less erroneous; but from the account I have given, on the best authority, of the flock at Rambouillet, it appears that they carry heavier fleeces of fine short wool than the Arlington breed do of long wool. I infer this from the letter of Mr. Foote, which is contained in the pamphlet: from his account five and a half pounds of wool, of about twelve inches in length, is the average of ewe lamb's wool in the first year, when the fleece is always the heaviest; because, instead of one year, it is generally of fourteen months growth. Mr. Lasteyrie, in his report to the National Institute in the year 1802, states, that the medium weight of full grown nursing ewes was eight pounds seven ounces; of the ewes of three years old which had no lambs, nine pounds thirteen ounces; and two-tenths ewes, ten and a half pounds. Now, making every allowance for the greater quantity of dirt contained in flocks kept as those in France are, I think we may state their weight as at least equal to those of Mr.

Foote's ewes; and yet Mr. Foote's sheep are evidently superior to the sheep of the country, whose average, under similar circumstances, would certainly not exceed three and a half pounds. My own, however, have not been so high as Mr. Lasteyrie's, and have not fallen much short of Mr. Foote's. Three full-bred ewes, all having lambs, gave the year before last eleven pounds and three quarters, or near four pounds each. Last year I did not keep a separate account, but as they were in better order, I think the average was near five pounds. This year seven fleeces, after they had been soaked twenty hours, and then washed in warm water, weighed twenty-six pounds, but this included two ram fleeces. Supposing them to have lost no more than common wool would have done, which, by such perfect washing, would not have been less than one third, they would then have weighed five pounds, which falls only half a pound short of the Arlington sheep. These seven fleeces would have sold for fifty-two dollars cash; whereas seven of Mr. Foote's fleeces, reduced one third by washing, would only have produced (selling at the usual price, thirty-six cents a pound) eighteen and a half dollars; and yet his sheep, being larger, would have demanded more keeping. If this is to be observed of one of the finest American flocks, how much greater will be the balance in favour of the Clermont Merinoes, when they are set in opposition to the sheep of the country? It is also an error to suppose that there is little consumption of fine cloth in this country. There are few people in our cities who wear such cloth as can be made from British wool, the finest of which will not make cloth of the value of more than thirteen shillings sterling per yard. The first, second and third cloths are all made from Merino wool of different grades of fineness. Nor, if we may believe Anderson, is there any cloth in which Merino and British wool are mixed; their qualities being so dissimilar, and they shrink so differently in the pulling, that they cannot be worked together. It is also a mistake to suppose that, with the same materials, we cannot make cloth of the same quality, and at the same price, with that of Britain. We now card by water, and spin with jennies; so that much of the labour is saved. I have for three years past been in the habit of manufacturing all the cloth necessary for my own use and for the use of my very large family; and I can say with certainty, that I can manufacture cloth of every quality from three to ten dollars per yard so much cheaper as to receive two dollars for my fine wool, and one dollar and fifty cents for the second, and at least one dollar for half-bred wool, and yet save 20 per cent. upon the manufacture, besides great gain in the superior strength of the cloth. This is very conceivable by those who calculate the expenses with which British manufactures are loaded before they come to us, which much more than compensate the difference in the price of labour. 1st. The manufacturer's profit. 2d. The purchase and transportation to a sea-port. 3d. The commission to the merchant in England. 4th. Four per cent. British duty. 5th. Freight. 6th. Insurance. 7th. American duty, seventeen and a half per cent. 8th. The merchant's profit, which is never less than fifteen per cent. 9th. The retailer's profit. Take all these items together, and they will not fall short of cent. per cent. The difference between the price of labour bestowed upon a piece of cloth of any degree of fineness in Europe or America, bears no proportion to this: for instance, two pounds of the finest Merino wool makes a yard of superfine broadcloth, which sells in England at twenty-four shillings sterling; the wool costs there fourteen shillings and six-pence sterling; the merchant's profit upon this is not less than ten per cent. or about two shillings and five-pence; the whole labour then employed in the manufacture is only seven shillings and one penny sterling. Suppose the price of labour here to be fifty per cent. higher, which exceeds the fact, then the price of the material being the same, the cost of making it here should be three-eighths more than in England, that is, three shillings and eight-pence upon twenty-four shillings; and cloth of that price, if manufactured in the United States should sell for twenty-seven shillings and eight-pence sterling, or about five dollars and sixty-three cents (five shillings sterling to the dollar), whereas no imported cloth of that quality can be purchased here for less than twelve dollars. What an immense saving then would it be to the United States to cultivate the breed of sheep which will furnish materials for an article on which they now pay upwards of 100 per cent.! what a field does it open both to the manufacturer and the farmer! While the one can afford to give two dollars and fifty cents per pound for wool, the other, even after having received that advanced price, can purchase his cloth much cheaper than he can now do, when he sells the fleeces of his flock at thirty-six cents per pound. But how much greater still will be the profit, if he manufactures his own wool into fine cloth for the market! I will venture to say, that cloth of ten dollars the yard may, in this way, be made superior in quality to British cloth, though perhaps not quite so well dressed, for three dollars per yard, of seven quarters wide, and give the farmer a profit of three dollars per pound for his wool, after allowing one dollar as a commission to the shop-keepers who sell his cloth. It is but justice, however, to the sheep I have mentioned, to wit, the Arlington breed, to observe, that their fleeces are adapted to purposes, to which those of the Merino cannot be applied with the same advantage; such as, the making of worsteds, camblets, sergers, and perhaps fine blankets. These manufactures require long combing wool, whereas cloth demands fine short wool, and one cannot be substituted for the other without loss. Wool which is intermediate is on that account inferior to either, as not being well adapted to cloth, and too short for combing. This is in some sort the character of the new Leicester or Bakewell wool; were it a few inches longer or shorter, it would sell much higher than it now does; its present price in the British market is only ten-pence sterling per pound, and yet it is of a tolerably fine staple.

As it is my wish to direct the choice of the

farmer to such sheep as will suit his wants, it will be proper to observe here, that if a farm is so circumstanced as to render it inconvenient to keep more sheep than will suffice to clothe the family and employ the leisure hours of the female part of it, I would recommend not to go beyond half-breed Merinoes. Whatever may be the stock of ewes, whether long or short-woolled, I can with certainty assert, that their lambs by Merino rams of the improved breed will carry heavier fleeces than the parent stock on either side. If they are short-woolled sheep, their fleeces will not only increase in quantity, but be much improved in quality: if they are long-woolled, the improvement will be more in the quantity and less in the quality. But in either case, the farmer, in addition to the increase of his wool, will find this essential advantage in crossing, that every fleece, if carefully sorted, will contain as much wool as will make cloth which no gentleman farmer need be ashamed to wear, and he will besides have different sorts, of inferior qualities, suited to his children and domesticks; but all will be more uniformly good than the wool of his old flock. Even if his flock consist of quarter-breed Merinoes, he will find an essential difference both in the quantity and quality of his wool. The average of my half-breed sheep is four pounds and three-quarters; whereas, with the same keeping, the stock from which the ewes came would not average more than three and a half; and among my half-breeds are many whose fleeces are so fine as to make cloth equal to imported cloth which sells at four dollars a yard. Sheep of this grade may be obtained at a very cheap rate by those who do not choose to go to the expense of a full-bred ram. Let them purchase a half-blood, in which they will have the advantage of a considerable choice of tups, and may select such as are best adapted to the flocks they wish to improve, either one that carries a large and long fleece, or one whose wool is short and fine. He will cost twelve dollars and fifty cents; the second year it will be easy to dispose of him for the first cost, and, by doubling the price, to purchase a three-fourths breed tup. This, with the quarter bred lambs, will at once give an half-bred flock at the expense of twelve dollars and fifty cents; he may then select the tups from his own flock, and sell his rams, and thus change his flock to half-blood without one cent expense, the fleeces of the rams overpaying the interest of the money and the keeping.

The extension of this valuable breed of

sheep is of great importance, as it relates to the community, the farmer, and the manufacturer, and even to every class of society, who are more or less affected by the cheapness and goodness of a fabrick which all employ, as well as by preserving a considerable capital within the State, and affording employment to numbers of indigent people, in whose happiness humanity interests itself. It therefore becomes the duty of the State to patronize and encourage them. This has been done with a laudable zeal by our Legislature: the premium upon the introduction of Merino rams has already had considerable effect; the bounty upon cloth will have a still greater; for it will soon be apparent to the people of every county, that the first prizes can only be taken by cloth made from the wool of the full-bred or mixed Merino sheep, and its superiority will be presented to their eyes in such form as to overcome the most inveterate prejudices. They will see that all attempts to make fine cloth from coarse wool is lost labour, and they will apply it to the use for which alone it is adapted. They will change their flocks as soon as possible, and as their wool will become more valuable, and meet with a ready market, they will find an advantage in increasing their flocks. Good policy however would dictate the continuance of this bounty for the term of at least ten years, that every man who has his flock still to change may have a prospect of benefiting by the liberality of the Legislature; for within less than half that time the competitors will be innumerable. In the mean time the premium on cloth operates indirectly as a bounty on wool; for many families that raise no wool, as the wives and daughters of mechanics settled in the country, or in villages, will find a pride and an interest in contending for the prize, and will become purchasers of the raw material at an advanced price: the most skilful weavers and dressers will be carefully sought out, and the celebrity they shall respectively acquire by having their names recorded with the prize cloth, will excite emulation among them, and afford full encouragement to those whose skill and industry shall best merit it. It may be a question how far it would be well to give a bounty upon certain articles made from the common wool of the country. For my own part, I believe it unnecessary, because all wool of that kind is already worked up in domestic manufactures, and is doubtless employed in that way which is most useful; if turned to a different use, perhaps neither the public or the individual will be so well served as they now are. But if a contrary sentiment should prevail, then I think the bounty should be given upon worsteds, serges, and blankets; because this would turn to its proper use that long wool which is misapplied in the making of cloth.

In Sweden the Merino sheep were introduced in 1723; they at once became a national object, and a bounty of twenty-five per cent. was paid upon the value of the wool to the grower, which was continued to 1781, when it was reduced to fifteen per cent. and in 1792 it was suppressed; Sweden then possessing upwards of 100,000 full-bred Merinoes, and being able to supply all her own wants without any importation from Spain; and what is very extraordinary, the sheep have undergone no change for the worse in the space of upwards of eighty years; though perhaps Sweden is of all the cultivated countries I know, least calculated for sheep; the length of the days during its short summer parches its barren fields, and for seven months it is buried in snow. They shear in July, and the average weight of a ewe's fleece, when washed, is three pounds. They keep up the Spanish practice of giving salt, particularly in wet weather.

To those who are unacquainted with Spanish

wool it may be proper to mention the manner in which it should be treated before they attempt to convert it into yarn. First, it should be carefully sorted; that on the neck, shoulder, back, and sides is the finest; that on the rump is almost equally as fine in the full-bred sheep, but not in the mixed breeds; the thighs and belly, the top of the head and forelock furnish a third sort: when sorted it should be put in a vat and pressed down, so as not to float when covered with water. In this state the vat should be filled with clean soft water, mixed with one-third of urine, and left to soak for about twelve or fifteen hours, or longer if the weather is cold; a cauldron is then put on the fire with a portion of soft water, and to this is added two-thirds of the water that covers the fleece: when it is so hot that the hand cannot bear it, the wool is to be taken in convenient parcels and put in an open basket, and, after the liquor is pressed out, conveyed to the cauldron, where it is washed in the basket, moving it about gently, so as not to twist it, for the space of two or three minutes; it is then suffered to drain into the cauldron, so as not to carry off the water; and when the whole is washed, it must be cleansed in running water: if the water in the cauldron gets too foul, it must be thrown away, and replenished with more of the liquor from the vat. This mode of washing preserves in the wool a certain portion of its grease, which makes it spin easier. When washed it may either be dried in the shade (the sun renders it harsh if too hot), or, what is better, it may be pressed in a cyder press, which dries it much quicker; when quite dry it should be laid upon cribbles, and beat with a bunch of rods, which softens it, and takes out a great proportion of the dust and hay-seeds; it is then picked carefully, not as common wool is, but by opening the flocks, which are in some measure tied together at the ends, and taking care not to break the wool to pieces. To fit it for spinning it should be greased with neat's-foot oil, and carded with cotton cards, wool cards being too coarse; and except the carding-mill is particularly fitted for it, and perfectly clean from common wool, it will run into knots, and be spoiled if carded at it. For domestic manufactures, from Spanish wool, I would therefore recommend the carding at home by hand. In Europe it it usual, before spinning, to wash the wool in the manner I have mentioned; yet, from some little essays that I have made upon the Merino wool, I am inclined to believe, if it is carefully picked, so as to leave no hay-seeds on the wool, and to open it perfectly before it is carded, that the finest thread may be made of unwashed wool: nor do I think that the yarn should be washed before it is wove; the grease adds to the strength, and renders it unnecessary to size the warp, as is usually done; more allowance should, however, be made for shrinking. This must not be considered as an ascertained fact, since I am now only in the course of trial.

As I do not write for professed manufacturers, but for farmers, many of whom have never given much attention to the best mode of fabricating cloths, I will venture to mention what, though well known to the first, the last may not yet be acquainted with. Common wool can hardly be too much carded, Merino may; the first gains by being broke to a certain degree, the last is injured. In spinning, the warp and the woof must be spun in contrary directions, because, as both open a little, and the object of fulling is to unite the ends of the wool, so as to raise the knap, if they untwisted the same way, they would unite less than if they met each other. This operation is effected by spinning the one with an open band, that is, a band that turns the spindle in the same direction with the wheel, the other with a cross band, which turns the spindle in a contrary direction. *Spectacle de la Nature* says it should be the warp, because this requires to be most twisted; the *Encyclopædists* say the contrary, assigning as a reason that the smoothest thread can be made by the open band, and that it is necessary that the warp should be particularly smooth and even; that inequalities are of less consequence in the woof, because they are corrected in the fulling. I have not sufficient knowledge of the subject to decide when such authorities differ: but in either case the woof must be spun as loose as possible. This renders the cloth soft, and makes it easier to raise the wool for shearing. To facilitate the spinning loose, a greater quantity of oil must be used in spinning the woof than in spinning the warp; for the first a fourth of the weight of the wool is used, for the last only one-eighth. This must be understood as applying only to fine yarn. Coarse thread is strong enough in itself to require but little grease. Olive oil is preferred in Europe; that which is most liquid must be used here. I will not venture a farther discussion on the subject of manufacturing cloth, since this information will be better acquired from practical manufacturers, who are to be found in almost every part of the State.

MISCELLANEOUS CHAPTER.

THE necessity I was under of sending the preceding part of this little treatise to Albany before the Legislature rose, though in an unfinished state, prevented my adding several matters that might properly have been digested into the body of the work, and which I must now either wholly omit, though they are useful to be known, as illustrative of several assertions which I have made therein, or throw them together in this indigested form. I have preferred the latter, and only pray that this chapter may be considered as a page from a memorandum book in which no order is to be expected.

Having been so fortunate as to excite the attention of my fellow citizens to the improvement of their breed of sheep, and this not only among experienced farmers, who need no instruction, but among a great number of those who have not heretofore felt interested in the subject, it is probable that, after having read what I present for their consideration, they will be inclined to look into the British writers, many of whom make use of technical terms well understood by those for whom they write, but which will need explanation here. I therefore state the following definitions of those most in use.

A male lamb, after he is weaned, is called, during the first year, a *Hog*, or *Hoggit*, a *Tag*. An ewe lamb, during the same period, is called an *Ewe Tag*, a *Gimmer*. In the second year the male is a *Shear Hog*, or a *two toothed Hog*, or *Tag*; the female a *Thaive*, a *Gimmer*, or *two toothed Ewe Tag*. Third and fourth year they are distinguished by the same names, with the addition of the number of teeth they have changed. The fifth year, having eight broad teeth, they are called *full-mouthed Sheep*. The age of the ram is generally denoted by the number of times they have been shorn; the first shearing being in their second year—*a shealing—one shear—two shear*, &c. In the north of England and in Scotland an ewe lamb, after weaning, is called a *Dimmont;* and in the west of England ram lambs are called *purlambs*. Tup and ram are synonymous terms for a covering ram. *Crone* signifies an old ewe.

Gestation.—The period for gestation in sheep is from twenty weeks to one hundred and fifty days: ewes will breed twice a year, and may even be made to breed thrice in that time, if they are kept in high order, and not suffered to nurse the lambs. I have two or three that have taken the ram this winter, since lambing, and one within eight days after lambing, though the lamb was running at her side. Whether the copulation will be productive remains to be seen; if it should (and I have no reason to think it will not), the lambs will fall in August; of course the ewes may take the ram in October, so as to lamb again in March, which would be three times in the course of the year. This, however, is a practice by no means to be recommended, as I think it would injure both the ewes and the lambs, and deteriorate the stock.

Lambs at birth.—In most breeds of sheep single lambs are more common than a greater number; but in some, as in the Dorsetshire, double lambs are nearly as usual as single. The Friesland and Tees-Water sheep, which are of the large long-woolled species, if well kept, bring from two to five lambs at a birth, and that sometimes twice in a year, if we may believe an old English writer, Barnaby Gage, who says, "It hath been seen in Guilderland, that five ewes "hath had in one year five and twenty lambs: "It may seem, peradventure, to many incredi-"ble, and yet no great marvel, since they have

" twice a year most times two, and sometimes " five at a time." Cully gives the following instance of fecundity in a Tees-Water ewe: When two years old she brought four lambs, then five, then two, then five, then two; the first nine within eleven months. The highest keeping is however necessary to cause this fertility.

Choice of rams.—I have already given directions for the choice of a ram, but as this is an object of much moment in forming a flock, it will be well to know the opinion of different agriculturalists. Columella recommends that the ram be tall, with a pendant woolly belly, a long tail, thick fleece, a broad forehead, twisted horns (though if without horns still better), and large testicles; not to be put to ewes till three years old, and not after eight.—Markham. " The ram large in every general part, with a " long body and large belly; forehead broad, " round, and well rising; a cheerful large eye, " straight short nostrils, a very small muzzle, " by no means any horns (for the hornless are " the best breeders, and do not endanger the " ewe as the horned do); a large upright neck, " somewhat bending, like the neck of a horse; " a very broad back, round buttocks; a thick " tail, and short jointed legs, small, clean, and

" nimble; his wool should be thick and deep, " covering his belly all over, also his face, even " to his nostrils, and so downward to his knees " and thighs." One would conclude from this description, that Markham, who wrote in the reign of Elizabeth, had copied from one of the Spanish rams imported by Edward IV. for no sheep in England answers to his model.

Folding.—I have passed over all observations on folding, because it is very little practised in this State; and wherever it is, it tends to the deterioration of the flock; and from experiments accurately made by Mr. L'Hommedieu, on Long-Island, it appears that the loss in wool, and injury to the sheep and lambs, exceeded the gain in manure.

Signs of health.—Signs of health in sheep are first a skittish briskness, clear azure eye, florid ruddy eye-strings and gums, teeth fast, breath sweet, nose and eyes dry, respiration free and regular, feet cool, dung substantial, wool fast and unbroken, skin of a fine florid red, particularly on the brisket. *Lawrence on Cattle,* p. 337.

Season of lambs falling, and food.—The season for putting the tups to the ewes depends upon the time it is wished to have the lambs fall; on that subject I have given my sentiments. The tups should be well fed in the season, barley and pease or Indian corn ground together should not be spared. I prefer myself to make them, or even wheat, corn, and rye, into bread, and give him a slice three times in the day. This is more convenient than giving loose grain, because, if your ram is as tame as he should always be, he will receive it from the shepherd's hand, so as not to render it necessary to take him up to feed, which is attended with a great deal of trouble. A little oats, or barley in troughs, or Indian corn scattered over the clear sward, from whence they will pick it up in single grains, will bring the ewes in heart (if they have no lambs) at any season that may be required. If the flock consists of aged ewes, with moderate care it will at least double annually. If a considerable proportion are ewes that have their first lambs, no care can prevent a loss of at least fifteen per cent. upon those of the young ewes, unless indeed the flock is very small. The Merino ewes are less prolific than those of our country, seldom producing twins.

Salt.—I have mentioned that salt was considered by the Spanish shepherds as essential to the health of sheep, and this sentiment is very general in every part of Europe except in England, whose situation renders the air sufficiently salt. The same consequence from similar causes takes place here. Upon Long-Island and elsewhere near the sea, the cattle require no salt, nor manifest any desire for it; whereas, on the north of the Highlands they eat it ravenously, and it is thought essential to their health. The ancients also entertained similar sentiments on this subject. Aristotle prescribed one peck every five days, during the summer, to one hundred sheep. We should consider this as a large allowance, but it would be readily eaten. They also observe, that however good your pastures may be, the sheep will tire of them if not changed, unless their appetites are kept up by salt.

I have been so often asked how much food is necessary for a sheep, and, indeed, a solution of the question is so important, that I think it right to state all that I can collect upon the subject. I have already given Daubenton's very accurate experiments, and they may serve as data to determine the comparative value of grass and other green provender. Lawrence, who appears to be an accurate and well informed agriculturalist, says, that a sheep will eat twenty pounds of turneps in twenty-four hours, but that one gallon of potatoes will generally suffice; from which it would follow, that less than eight pounds of potatoes are equal in value to twenty pounds of turneps. As the size of the sheep are not given, we are ignorant to what species of them to refer this assertion; and yet they differ very widely from each other in their size and form, upon both of which the quantity of food necessary for their support must in a great measure depend. He states also the comparative quantity of food required by a sheep and an ox as eight or nine to one. A course of experiments was made to determine the relative quantity of food eaten by different kinds of sheep. Four of the South Down breed, whose weight is about equal to twenty pounds a quarter, eat in seven days twenty-nine pounds of cabbage and seventy pounds of hay. This comes to two and a half pounds of hay, and one pound nine-penny weight of cabbage, which exceeds Daubenton's calculation; but not more than may be accounted for from the different size of the sheep. What follows is astonishing: the same sheep eat daily of green vetches one hundred and seven pounds, or twenty-six pounds per diem each. Vetches must by this be less nutricious than natural grass, nearly as two and a half to one. It is possible that clover would afford a similar result. This ought to be investigated. It is not less important to know the number of Merino

sheep that may be kept upon an acre during the summer. It appears from Lord Somerville's experiments, that Ryeland ewes, crossed by Merino rams, produced a fine stock of wethers, which were fit for the butcher at two years old, and weighed from fifteen to twenty pounds the quarter, and tallowed well. He adds, that they may be stocked hard, as the same land which carried indifferently forty-five long-woolled ewes, maintained, in good plight, one hundred and fifty Ryelands, the lambs of which were weaned in high order. These lambs were summered on the same land, at more than twelve per acre; and although kept hard during the winter, the wethers fatted to sixteen pounds the quarter. Lawrence adds, "I have been assur-
" ed, from good authority, that 221 acres of
" pasture vetches and turneps, being the whole
" of the land on which the sheep run last year,
" the profit of a flock of this breed (half blood
" Merinoes) amounted to £1592 9s. 2d. ster-
" ling; but working oxen, and other horned
" cattle ran over the same land, for which must
" be deducted 31 acres, so that there remain
" but 190 acres chargeable to the sheep, mak-
" ing a return of £7 4s. sterling per acre in a
" most disadvantageous season, on account of
" the drought. In general land worth a gui-

" nea and a half per acre, will carry and keep " in good store state six and a half Spanish Rye-" lands, from four to four and a half Spanish " South Downs, allowing turneps, pease, and " haum in the dead winter months. The larg-" est breed of South Down are stocked in Sus-" sex, at the rate of four per acre; of full-bred " Merinoes, an acre will carry a proportionably " greater number. They have been found very " apt to take on both flesh and fat; for two " ewes exhibited at Bath fifty guineas were re-" fused. One acre of good grass will keep 500 " couple a day. The harder you stock, the " more grass, and the more sheep you may " keep; besides that, hard stocking will render " the coarse grass fine."

I have asserted that a cross with Merino rams upon any stock would add at least one-third to the value of the wool, taking quantity and quality together; I, however, presumed this from my own experience, and from a cross I had seen between a Merino and a long-woolled British sheep. I have since met with the following confirmations of my assertions, even carrying them much farther than I have done. I wished to confine myself to the strictest bounds, that every man's experiments might at least verify mine, and excite that confidence in my recommendations which I wish to inspire. "Mr. Arbuthnot, of Bath (formerly a woollen "manufacturer), is fully convinced of the prac- "ticability of equalling Spanish wool in Eng- "land; he has for several years tried the Spanish "cross with the Wiltshire breed, *nearly trebling* "the wool in quantity, and improving it ex- "tremely in quality. The carcasses are reduc- "ed in size, but improved in the mould, and "the disposition to fatten increased." I quote this from Lawrence, who also adds, that in England Mr. Tollet had gone very successfully into the breeding of Merinoes, getting his stock from Lord Somerville. He mentions a tup of his that was adjudged, at thirteen months old, to weigh twenty pounds the quarter, for whom he refused two hundred guineas, and one hundred for his hire for a season. He then states the effect of his crosses, in the first degree, on South Down and Ryeland ewes, and makes the average increased value eleven shillings and six-pence sterling per fleece, or 150 per cent. advance upon the wool on *one cross only*.

The following opinion of Mr. Tollet merits great attention. He judges that an acre of land which will keep three South Down sheep, similar to our best sheep, would be sufficient to keep four Merinoes. The produce in wool of the South Downs would be thirteen shillings and six-pence per acre, that of the Merinoes £3 15s. 6d. sterling.

The advantages of the introduction of the Merino sheep have even been acknowledged in the southern hemisphere. In New-South-Wales they have been bred to great advantage, and the wool has not degenerated. Captain M'Arthur states, that in the year 1801 there were 6720, a few having been originally introduced from the Cape of Good-Hope, to which place they had been transported by the Dutch. He exhibited to the Secretary of State the fleece of a coarse ewe, valued at nine-pence per pound, and that of her lamb by a Merino ram, valued at three shillings sterling.

It is unnecessary to multiply proofs on this head; enough has been said to convince every unprejudiced man that his profits from sheep may be doubled by changing his stock. If he has the means, let him procure full-bred rams; if he has not, let him take others of inferior grade, of which many may now be procured: he will by this be approximating the great object he has in view; and let him be assured, that even one-quarter Merino blood will greatly improve his stock the very first year, even to double and treble the amount of his advances:

besides this, it will lay the foundation of a good stock to breed upon when he is enabled to procure rams of a higher grade. The mutton of the Merino sheep is acknowledged, by a variety of writers whom it would be useless to quote, to be of very superior quality, and easily fatted. Of this fact, so far as it relates to the full-bred, I have no experience; the half-bred wethers which I have fatted for my own table were certainly not inferior to the country breed, either in size, fat, or flavour; they weighed sixteen pounds the quarter, and tallowed well.

I have, upon a former occasion, given an account of the British sheep. As the attention of our farmers has been more fixed on them than on those of any other country, it will not be improper to offer the following short recapitulation of them by the celebrated British farmer Cully, with some additions by Lawrence, taken from the book of the latter.

Concise Description of British Sheep.

Name.	Horns.	Colour.	Quality of wool.	Weight of fleece.	Price of wool per pound.		Wether per quarter.	Age when killed.
No. 1. Tees-Water	No horns	White legs and face	Long wool	9 lb.	1s.	0d.	30 lb.	2 years
2. Lincolnshire	Do.	Do.	Do.	11	1	0	25	2
3. New Leicestershire	Do.	Do.	Do. fine	8	1	0	22	2
4. Cartwold	Do.	Do.	Do. do.	9	1	0	24	2
5. Romney Marsh	Do.	Do.	Do. do.	8	1	0	22	2
6. Dartmoor or Bampton	Do.	Do.	Do.	9	1	0	25	2
7. Exmore	Horned	Do.	Do. coarse	6	0	10	16	2½
8. Heath	Do.	Black face and legs	Do.	3	0	8	15	3½
9. Hereford, Ryeland	No horns	White face and legs	Short fine wool	2½	2	9	14	3½
10. Morf, Shropshire	Horned	Black and speckled	Do.	1¾	3	0	12	3½
11. Dorsetshire	Do.	White do.	Do.	3½	1	6	18	3½
12. Wiltshire	Do.	Do.	Short middling	3	1	6	20	3
13. Berkshire	No horns	Black and white	Long wool	7	0	10	18	2½
14. South Down	Do.	Speckled do.	Short wool	2½	2	3	18	2
15. Norfolk	Horned	Black do.	Do.	2	1	8	18	3½
16. Hardwick	No horns	Speckled do.	Do.	2	0	8	10	4½
17. Cheviot	Do.	White face and legs	Do.	3	1	1	16	4½
18. Dun faced	Do.	Dun do.	Do.	1½	3	2	7	4½
19. Shetland	Do.	Various colours	Do. fine cottony	1¾	3	2	8	4½
20. Spanish	Rams horned	White	Do. superfine	3½	6	6	14	2½
21. Spanish cross			Do. fine	2¾	3	2	16	2

It will also be of use to be acquainted with the several breeds of Spain, as a direction to those who may endeavour to import sheep from thence; for, though every species of the Merino is valuable, yet they differ widely from each other in beauty, in form, and in fineness of fleece, as may be judged from the prices in Spain, where Leon and Escureal wool sells for 100 cents, while that of Aragon brings only 60; with several intermediate grades, which I have given in a former communication to the Society for Useful Arts. Those most noted are the sheep of the *Escureal,* of *Guadaloupe, Paular,* of the Duke *D'Enfantado, Monturio,* and of the *Negretti.* The first, for fineness of wool, is the most perfect of all the travelling flocks of Spain; the second, for form, fineness and abundance of the fleece; the third (Paular), with similar fleeces, are longer bodied. The lambs of this stock, and that of the Duke D'Enfantado, are commonly dropped with a thick covering, which changes into very fine wool. The Negretti are the largest breed in Spain. It is from the last stock that England has drawn he Merinoes.

APPENDIX.

IT would be waste of time to speak here of the long list of maladies which attend sheep in Europe, most of which I believe are to be attributed to injudicious treatment, particularly in folding and over-driving them. I shall confine myself, therefore, to those only which I have witnessed in this State, the number of which is very small.

Pinning and scouring.—Lambs, soon after the birth, are subject to a disorder called pinning. It consists in the excrements being so glutinous as to fix the tail to the vent, which, if neglected, will often kill the lamb. The remedy is to wash them clean, and to rub the buttocks and tail with dry clay, which will prevent any further adhesion. Lambs are also subject to scouring, or purging. This generally arises from being kept too cold; sometimes from the quality of the ewe's milk. They should, with the parent ewes, be put into a warm, dry, sheltered cot: the ewes should have plenty of nutricious food given them; such as oats, old Indian corn, and wheat bread: care should be taken that they nurse their lambs duly, for it often happens that this complaint is aggravated by a penury of milk; in this case the deficiency should be supplied by cow's milk boiled, or by letting the lamb suck a cow.

Hove.—Sheep turned into clover too suddenly, and with empty stomachs, are sometimes inflated by the wind in that organ, the orifice of which is stopped by the food they have taken. This the farmer calls being hove. All ruminating cattle are subject to it. On being affected with it, they swell very suddenly, and, unless speedily relieved, they die. Several remedies are prescribed for this disease; the first and most effectual is to plunge a knife into the paunch. The sheep will swell most on the left side, and a part of the swelling will be very protuberant below the hip-bone. Into this protuberance plunge a knife, sharp at the point and dull on the edge, so as not to cut unnecessarily sideways. The depth must be regulated by the degree of swelling; there is little danger of going too deep, and the knife must enter the stomach to be effectual. The aperture must be kept open till all the wind is discharged, which will be in a few minutes. Another remedy is to take a piece of rattan or grape-vine, with a natural or artificial knob at the end, covered with cloth or leather, and to thrust it down the throat into the stomach. This will open the aperture, and the wind will be discharged. To these surgical operations chemical remedies are sometimes substituted, and should be first tried if there is time for it; a pint of linseed oil has been successfully given, or a solution of potash or common ley: both of these will combine with the carbonic acid in the stomach, and may, of course, effect a cure if given in sufficient quantities to absorb the air.

Purging.—Sheep turned into pastures in the spring are very subject to a purging, principally from a change in diet, and laying on the wet ground after being turned out of their dry folds. This is in general a malady of little consequence, and perhaps is salutary upon the whole, if not too great or too long continued. I have never used any remedies for it; but I conceive that folding in their winter cot, upon dry litter, for a few nights, with a handful of hay and grain, would check it: to this may be added salt, mixed with any absorbent earth, which the sheep will eat very readily. If any are so much affected as to be weakened by it, and the disorder does not yield to those remedies, a dose of castor oil, and housing, with dry food, particularly a crust of wheat bread, will generally restore them.

Scab.—The disorder which most affects our sheep is the scab. This appears first by the sheep rubbing themselves and pulling out their wool. As soon as this is observed, or when loose locks of wool appear to rise upon their backs and shoulders, they should be examined, the wool taken out, and a little spirits of turpentine and hog's-fat rubbed on the place. If this be neglected for some time, and the disorder increases, the skin will feel hot and hard to the hand; and, if longer neglected, the wool will pull off in large quantities, and the scab be converted into a sore, from which a small quantity of matter will ooze and clot the lower part of the wool; and if altogether left to itself, the whole fleece will drop off, and the sheep pine away; but they will generally, in a certain degree, recover from the first attack when they get to grass; they will, however, be very liable to take it again the next winter, and then it generally proves fatal. I have never failed to cure mine in ten days by the following treatment. First, I separate the sheep (for it is very infectious); I then cut off the wool as far as the skin feels hard to the finger; the scab is then washed with soap-suds, and rubbed hard with a shoe-brush, so as to cleanse and break the scab. I always keep for this use a decoction of tobacco, to which I add one-third by measure of the ley of wood ashes, as much hog's-lard as will be dissolved by the ley, a small quantity of tar from the tar-bucket, which contains grease, and about one-eight of the whole by measure of spirits of turpentine. This liquor is rubbed upon the part infected, and spread to a little distance round it, in three washings, with an interval of three days each. I have never failed in this way to effect a cure when the disorder was only partial. By attention I have always prevented its attaining so great a degree of malignity as to suffer the sheep to lose more than eight or nine inches square of its wool; I cannot, therefore, say whether it would cure a sheep infected so as to lose half its fleece, in which state I have seen many flocks. In such case I think recourse should be had to mercurial ointment, which has been strongly recommended by Sir Joseph Banks; who says that it is a very safe remedy if applied with care. He directs the wool to be opened, and a streak to be made down the back, and from thence down the thighs and ribs. A good shepherd will, however, prevent its ever attaining to this degree of malignity. Daubenton recommends spirits of turpentine and hog's-lard, or suet without any other mixture, as less hurtful to the wool, and equally effectual. The fine-woolled sheep, and particularly the rams that are exhausted by covering, are most subject to it; and in the fine-woolled flocks it is also most difficult to cure. It spreads not merely by contact of one sheep with another, but by their laying upon the same ground, or rubbing against the same post.

Staggers, Dizziness, &c.—This disorder I have already described, and have detailed my success in curing it by patience and attention. It seldom seizes sheep of more than one year old, and is generally considered as incurable; though some affect to cure it by trepanning, or by running a sharp wire up the nostrils into the brain, so as to discharge through the nostrils the water which is collected there, and which is said to be the cause of the disorder. When it is seated near the upper part of the brain, it may be distinguished; the skull bone becoming soft immediately about it, so as to yield to the pressure of the finger. But having no other experience on the subject than that which I have mentioned, I can say nothing as to the efficacy of the harsh remedies proposed.

Pelt-Rot.—This is often mistaken for the scab, but is in fact a different and less dangerous disease: in this the wool will fall off, and leave the sheep nearly naked; but it is attended with no soreness, though a white crust will cover the skin from which the wool has dropped. It generally arises from hard keeping and much exposure to cold and wet, and, in fact, the animal often dies in severe weather from the cold it suffers by the loss of its coat. The remedy is full feeding, a warm stall, and anointing the hard part of the skin with tar, oil, and butter.

Tick.—This insect is extremely hurtful to sheep; it often reduces their flesh by the pain it induces, and spoils their wool by their tangling and rubbing it against trees and fences. Lean sheep are frequently so covered by them as to occasion their death. The remedies applied in England are solutions of arsenick or corrosive sublimate, and decoctions of tobacco. The first are dangerous to the operator, and may occasion fatal accidents; the last are hurtful to the sheep, if not carefully applied; but all are ineffectual on thick-woolled sheep, because it is impossible to diffuse them equally.

I have happily discovered a mode of entirely destroying the tick, which is easy in the application, and attended with no danger. Take a bellows, to the nozel of which a pipe must be affixed capable of containing a handful of tobacco; (the refuse from the tobacconist's will answer); set fire to the tobacco, and while one man holds the sheep between his knees, let another open the wool, while a third blows the smoke into the fleece; close the wool on the smoke, and open another place a few inches from it, and so go over the whole sheep, blowing also under the belly and between the legs: in twenty-four hours every tick will be killed. The whole operation may be performed upon a sheep in about two minutes.

Cold, and its consequence.—When sheep are very ill kept, or when they lay upon damp or wet ground in the spring and autumn, they are subject to colds, which appear by the discharge of mucus from the nose and eyes, and sometimes by blindness. The cure is warmth, dry litter, and good food. It will however happen, that some sheep have at all times this discharge from the nose; but, upon examination, those will generally be found to be old, and should be fatted as soon as possible, as they disfigure a flock, and do not pay for their keeping.

Dogs.—This is one of the severest maladies under which our sheep labour; it generally attacks a whole flock suddenly, in which they run from each other in every direction; their wool and flesh appear to be torn to pieces; many, when the disorder is seated on the throat and neck, die suddenly; others appear to be wounded in different parts of their bodies, and die in great torment. Sometimes the greater part of a flock are carried off by it in one night, and the expense and trouble incurred for years in raising a fine flock are instantaneously destroyed; for such is the nature of this complaint, that no attention on the part of the owner can prevent it. The remedy is good wholesome laws, steadily persisted in—firmness in the magistracy in carrying them into effect—sufficient good sense in the people to aid in enforcing them, a readiness to respect the property of their neighbours, and to sacrifice boyish attachments to the general interest of the community.

Method of bleeding sheep.—In inflammatory disorders bleeding may be necessary. This is performed by cutting the ear, or the tail, or in the temple. The first and last do not yield much blood, and cutting the tail leaves a considerable wound. Daubenton recommends bleeding in the lower part of the cheek, at the spot where the root of the fourth tooth is placed, which is the thickest part of the cheek, and is marked on the external surface of the bone of the upper jaw by a tubercle sufficiently prominent to be very sensible to the finger when the skin of the cheek is touched. This tubercle is a certain index to the angular vein which is placed below; and this vein extends from the under border of the jaw beneath near its angle, to below the tubercle, which is seated at the root of the fourth cheek tooth; farther the vein bends and extends to the cavity of the eye-brow. The shepherd takes the sheep between his legs; his left hand, more advanced than his right, which he places under the head, and grasps the under jaw near to the hinder extremity, in order to press the angular vein, which passes in that place, to make it swell; he touches the right cheek at the spot nearly equidistant from the eye and the mouth, and there finds the tubercle which is to guide him, and also feels the angular vein swelled below this tubercle; he then makes the incision from below upward, half an inch in length below the middle of the projection which serves to guide him.

The following table, designating the time that it will take to form a Merino flock from one hundred common ewes, has been composed and transmitted to me by Simeon De Witt, Esq. Surveyor-General of the State, since this essay went to the press. It is very ingenious, and carries with it such a strong conviction of the practicability of changing the whole stock of the State into Merino sheep in the course of a very few years, that I am sure my readers will examine it with pleasure. One reflection will occur, to wit, that it supposes the ewes of the first year to breed. This will happen if they come early and are well kept; if otherwise, many of them will not drop lambs the first year, and indeed many good farmers are of opinion that they should not take the ram till they are eighteen months old. The table furnishes the means of calculating the stock of full-blood at the end of eight years in either case. Mr. De Witt also supposes only eighty lambs from one hundred ewes, though in general there will be one hundred lambs raised from that number of ewes, if they are properly kept; the double lambs making up for those that are lost. It would be curious to follow it through all its ramifications, and state the number of Merinoes of different degrees of blood, that would originate in this flock from sheep sold within the eight years, and to calculate the profits that resulted from the change during its whole progress by the sale of the males and the increased value of the wool.

Scheme for transmuting a flock of 100 *common Ewes and their issue into Merino Sheep.*

Years.		Common Ewes.	$\frac{1}{2}$ bld. *Rm.*	*Ew.*	$\frac{3}{4}$ bld. *Rm.*	*Ew.*	$\frac{7}{8}$ bld. *Rm*	*Ew.*	$\frac{15}{16}$ bd. *Rm.*	*Ew.*	$\frac{31}{32}$ bd. *Rm.*	*Ew.*	$\frac{63}{64}$ bd. *Rm.*	*Ew.*	No. of flock.
	Sold	a	B	b											
1	B	100	40	40											140
			C	c	D	d									
2	CD		40	40	16	16									196
			E	e	F	f	G	g							
3	EFG$\frac{1}{2}$a		20	20	32	32	6	6							204
					H	h	I	i	K	k					
4	HIK$\frac{1}{2}$a				40	40	19	19	2	2					215
					L	l	M	m	N	n	O	o			
5	LMNOb$\frac{1}{2}$c				16	16	35	35	10	10	1	1			217
							P	p	Q	q	R	r			
6	PQR$\frac{1}{2}$cde						35	35	24	24	5	5			225
							S	s	T	t	U	u	V	v	
7	STUVf						22	22	38	38	14	14	2	2	269
									W	w	X	x	Y	y	
8	WXYhlg								44	44	29	29	8	8	288

This scheme supposes one to commence the establishment of a Merino flock with two Merino rams and one hundred common ewes, and that one hundred ewes will annually bring forty ram and forty ewe lambs. The 3d, 4th, 5th,

6th, 7th, and 8th columns show the number of each grade produced in the opposite noted in the first column, and the last column the number of the flock, after selling those referred to by the letters in the second column. The ram lambs are supposed to be every year sold; the number given for the flock are therefore of ewes only. After the seventh or eight year the flock of full-blooded sheep may soon be increased to any number required for a farm. The fifteen-sixteenths, and succeeding grades, are considered of the full-blood.

Having had occasion frequently to mention the Clermont Merinoes, I thought it might be satisfactory to give an account of the last sheep-shearing, in order to show, that in all I have said of them I have rather under than overrated their produce. The prices set down in the following table are those at which the wool actually sold unwashed.

24

No. of sheep.	Quantity of wool.		Average of each.		Price sold at.	Value of fleece per head.	Total amount.	Profit and loss, deducting one dollar and fifty cents for keeping per head.	
29 common	114 lb.	4 oz.	3 lb.	15 oz.	$ 37½	$ 1 47	42 84	Loss on each fleece	$ 3
83 ½ bred ewes	393	7	4	11	75	3 53	295 7	Gain on each fleece	2 3
47 ½ bred wethers	236	4	5	6	75	4 11	177 19	Do. do. do.	2 55
30 ¾ bred ewes	156	6	5		1 25	6 25	195 47	Do. do. do.	4 75
3 ¾ bred wethers	16	6	5	7	1 25	6 83	20 47	Do. do. do.	5 25
27 ⅞ bred ewes	139		5	2	1 50	7 69	208 50	Do. do. do.	6 81
7 full-bred ewes	36		5	2	2	10 25	72	Do. do. do.	8 75
Full bred rams.									
Clermont, 14 months old	9	6	9	6	2	18 75	18 75	Do. do. do.	17 25
Rambouillet	9		9		2	18	18	Do. do. do.	16 50
Columbus and Hornless	12	14	6	7	2	12 85	25 70	Do. do. do,	11 35
							$ 1073 99		

Weight of Rams.

Rambouillet	140 lb.
Clermont	126
Columbus	123
Hornless	122